SECOND EDITION

STUDY GUIDE TO ACCOMPANY
PHYSICAL GEOGRAPHY

RALPH C. SCOTT
Towson State University

Prepared by

RICHARD L. HAIMAN
California State University, Chico

PAUL Z. MELCON
California State University, Chico

West Publishing Company
St. Paul   New York   Los Angeles   San Francisco

TABLE OF CONTENTS

PREFACE

Welcome to this study guide for the second revised edition of your introductory physical geography text by Professor Scott. This text is eminently readable and is much more "student friendly" than most texts on the market. It sequentially moves through a survey of fundamental topics in physical geography. It is concise and clear; difficult material is accompanied by photo-graphs, figures, and tables which assist in explaining the material. Special sections at the end of chapters focus on particular applications of the material covered in the chapter.

This study guide is intended to be used in conjunction with both the textbook and lecture materials in the course. The text should be read initially to acquaint yourself with the terminology and general concepts in the chapter. Each chapter of the study guide begins with a series of learning objec-tives, followed by a comprehensive listing of key terms and concepts for the chapter. A self-testing section follows; we have chosen to include a mixture of true/false, matching, multiple choice, short answer, and essay questions because these are the preferred formats for university examinations. Note that the listing of question types moves from the least difficult to the most difficult; to best evaluate your knowledge, be certain to attempt the essay questions before proceeding to the next chapter.

Before beginning a thorough reading of the text chapter, the learning Objectives in the study guide should be read. At this time the text should then be re-examined in conjunction with the study guide, completing short responses for the Key Terms and Concepts. Having completed the Key Terms and Concepts section, you are now ready for an initial period of self-testing. First complete the true/false and matching sections for each chapter. Turn to the answer key and correct your work. If you missed an answer, refer to the text for the correct response. At this time, complete the multiple choice questions for each chapter, again checking the key and referring to the text as necessary. Finally, outline brief responses to the short answer and essay sections.

The last part of each section in the study guide is a place location exercise keyed to a continent or other major region of the world. Part of geography is knowing the location of places; this is similar to understanding the abbreviations of elements in chemistry, or knowing dates in history. Use an atlas to locate the place names in the list for each map. Global geographic literacy begins with knowing where places are!

We hope that this study guide is useful in your work. Physical geography is a fascinating subject. If you have any comments on the study guide, please send them to us (Department of Geography, California State University, Chico, Chico, CA 95929-0425).

We have enjoyed the process of preparing this study guide. This work would not have been possible without the assistance of Mr. Charles Nelson, cartographic technician of the Geography Department, and Ms. Jennifer Robison, department secretary, who typed the manuscript and developed the final format for this study guide.

Good luck and our best wishes as you pursue the study of physical geography.

Richard L. Haiman, Ph.D.
Professor of Geography

Paul Z. Melcon, Ph.D.
Associate Professor of Geography

California State University, Chico

C H A P T E R   1

THE SCOPE OF GEOGRAPHY

LEARNING OBJECTIVES

Knowledge is generally organized on a topical basis; botanists study plants, sociologists study human society; geologists study the subsurface materials of the earth. Geography is not a topical discipline; instead, it is concerned with locational distributions. Explaining why a phenomenon on the earth's surface occurs in a particular place is the focus of geographical inquiry. Geography reveals itself not by the topic investigated, but rather by the question asked.

Geographic study attempts to explain the order and disorder of phenomena on the earth's surface. It broadly examines physical and human components, while realizing that interrelationships prevent total separation of contributing factors. As you continue in this course, remember that geography attempts to provide an explanation for the natural and human features that you see about you on a daily basis. Geography explains the physical and cultural landscapes of your everyday life.

KEY TERMS AND CONCEPTS

Provide a short definition or description of the following key words and concepts from the chapter.

<u>Geography</u>

    Place

    Spatial Interaction

    Locational Distributions

<u>Geographic Subdivisions</u>

    Physical Geography

        Atmosphere

        Hydrosphere

        Lithosphere

        Biosphere

    Human Geography

<u>Geographic Themes</u>

    Region

        Generalization

        Boundaries

    Change Over Time

    System

        Closed System

        Open System

<u>Types of Analysis</u>

    Systematic Approach

    Historical Approach

    Geographical Approach

TESTING: TRUE/FALSE AND MATCHING

1. _____ Geography is a topical discipline.

2. _____ Geographers study earth processes and their effects on humanity.

3. _____ Human geography is a subdiscipline of physical geography.

4. _____ Botany is similar to geography because both are organized by the type of thing to be studied.

5. _____ Immanuel Kant was a famous German geographer of the late eighteenth century.

6. _____ Geography is both a natural science and a social science.

7. _____ Interrelationships are infrequently studied by geographers.

8. _____ A geographer is more likely to see a forest than the trees within the forest.

9. _____ Integration of knowledge is not a primary goal of geographers.

10. _____ The subject matter of geography is difficult to encounter in the world around us.

11. _____ The subdisciplines of geography are more specialized than the discipline as a whole.

12. _____ Soils are of no interest to physical geographers.

13. _____ Human geographers focus on distributions controlled by people.

14. _____ Place refers to the geographic location (latitude and longitude) of a spot on the earth's surface.

15. _____ Regional analysis is a means of focussing on the characteristics of small areas.

16. _____ The boundary lines on maps generally correspond to lines found on the surface of the earth.

17. _____ In this text, the concept of a system is intimately concerned with the availability and the movement of energy.

18. _____ "Geographic literacy" deals only with place names on maps.

19. _____ A useful means of examining the subsystems of the earth is by designating "environmental spheres."

20. _____ Geography is a broad study of earth phenomena organized on the general theme of spatial location and interaction.

TESTING: MULTIPLE CHOICE

Choose the best response for each of the following multiple choice questions. Each question has only one correct answer.

21. Geography is defined by:
    a.   its subject matter.
    b.   its concern with Kant's philosophy of the organization of knowledge.
    c.   its intensive examination of minerals and rock types.
    d.   its method of approach.

22. Which of the following is not one of Kant's major subdivisions of approaches to knowledge?
    a.   historical approach
    b.   geographic approach
    c.   segregative approach
    d.   systematic approach

23. Which of the following is not a topically oriented discipline?
    a.   history
    b.   biology
    c.   chemistry
    d.   psychology

24. A central belief of geographers is that:
    a.   the distributions of phenomena on the earth's surface are not random.
    b.   only those distributions which are unrelated to human beings can be properly studied by geographers.
    c.   geographers study the minute details of location which are of no interest in daily lives.
    d.   the lack of an integrative approach to inquiry is a strength of the discipline.

25. Which type of scientist would be most likely to go to the top of a hill and write down everything in her sight?
    a. geologist
    b. geographer
    c. chemist
    d. art historian

26. Place includes:
    a. only the map location of a spot
    b. only the human components of the landscape surrounding a spot
    c. only the physical components of the landscape surrounding a spot
    d. every bit of possible information about a spot

27. The familiar expression "unable to see the forest for the trees" implicitly recognizes:
    a. a common pitfall associated with the systematic approach to knowledge.
    b. a type of error which results from focused study of detail.
    c. the need for individuals trained to look at problems from a broad perspective.
    d. each of the above answers is correct.

28. Which are the two key concepts of geography?
    a. distributions and interrelationships of phenomena
    b. location and physical site characteristics
    c. location and place
    d. causes and consequences

29. Which of the following is not a topically defined subdiscipline of physical geography?
    a. biogeography (the study of plant distributions)
    b. geomorphology (the study of landforms)
    c. climatology (the study of climates)
    d. economic geography (the study of economic patterns)

30. Which of the following is not a topic discussed as an area of human geography?
    a. transportation routing
    b. agricultural activity
    c. soil type
    d. recreational planning

31. The essential difference between human geography and
     physical geography
     a.   is the focus of the latter on factors controlled or
          influenced by human beings.
     b.   is the focus of the former on factors associated with
          natural earth phenomena.
     c.   is emphasis of physical geographers on "environmental
          spheres" while human geographers commonly ignore the
          natural components of landscapes.
     d.   none of the above answers is correct.

32. Unlimited energy supply:
     a.   is not characteristic of the earth system.
     b.   is basic to an open system.
     c.   is available to all organisms which are members of
          closed systems.
     d.   is an essential characteristic of closed systems.

33. Which of the following is not an "environmental sphere"?
     a.   atmosphere
     b.   hydrosphere
     c.   biosphere
     d.   morphosphere

TESTING: SHORT ANSWER AND ESSAY

1. How is it possible that geography is both a natural
    science and a social science?

2. What are the basic differences between a topically
    oriented discipline and one which studies everything from
    a single perspective?

3. What are the basic differences between a topically
    oriented discipline and one which involves the study of
    everything from a single perspective?

4. What are the sources of energy for the earth's systems? Which is the most important source of energy? Why is systems analysis commonly used for geographic studies?

5. Draw a tree structure which shows the characteristics which combine to make up the basic characteristics of a spot on the earth's surface. (Hint: start with the heading geography; for the initial subheadings use physical geography and human geography; for further subdivisions of physical use litho-, bio-, hydro-, and atmos; under these subtopics list examples of specific place characteristics.

PLACE AND PHYSICAL FEATURE LOCATION

WORLD MAP EXERCISE

Locate the following by name or number on the map of the **World** (NOTE: you may wish to sketch in the major landforms).

GEOGRAPHIC REFERENCE LINES

1. Antarctic Circle 66 1/2^{O}S
2. Arctic Circle 66 1/2^{O}N
3. Equator 0° N,S
4. Tropic of Cancer 23 1/2^{O}N
5. Tropic of Capricorn 23 1/2^{O}S
6. Prime Meridian 0° E,W
7. International Date Line 180° E,W
8. North Pole 90°N
9. South Pole 90°S

CONTINENTS

10. Africa
11. Antarctica
12. Australia
13. Asia
14. Europe
15. North America
16. South America

OCEANS

17. Arctic Ocean
18. Indian Ocean
19. North Atlantic Ocean
20. North Pacific Ocean
21. South Atlantic Ocean
22. South Pacific Ocean

HEMISPHERES

23. Northern
24. Southern
25. Eastern
26. Western

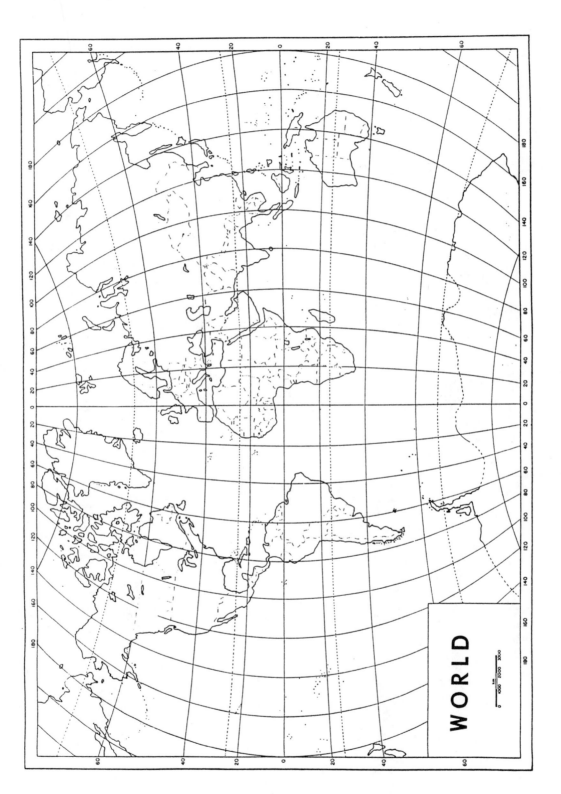

WORLD

C H A P T E R 2

THE PLANETARY SETTING

LEARNING OBJECTIVES

1. Develop a familiarity with the basic geometry of the solar system and the terminology associated with it.

2. Become familiar with the coordinate system used for locating features on the earth's surface.

3. Investigate the large and small scale motions of the earth through space and become familiar with how they alter our environment.

4. Add the specific terms to your vocabulary which are used to describe special positions in the earth's revolution about the sun.

5. Understand why the tilt of the rotational axis is critical to seasonal changes at the earth's surface.

6. Increase your awareness of the shape of the earth and the relative importance of the departures from sphericity.

7. Learn proper usage of compass directions, the difference between magnetic directions and true directions.

8. Determine the particular aspects of latitude and longitude which are commonly used in describing locations on the earth. This will provide a needed frame of reference for subsequent discussions.

9.   Investigate how global time is organized into a system
     which, while taking into account the local position of
     the sun, provides for worldwide ordering of time zones
     and the date.

10. Become familiar with the standard time zones, converting
     time to a common reference, and the political manipula-
     tion of time zones.

11. Gain a complete understanding of the role of the
     International Date Line in maintaining the integrity of
     the world time zones. Gain an appreciation of why the
     date must change as this line is crossed, even while the
     time of day remains constant.

12. Continue the study of time with an analysis of the origin
     of Daylight Saving Time and the effect it has on our
     lives.

13. Introduce the concept of the seasons and gain a
     fundamental understanding of the causes of seasonal
     changes on the earth. These changes must be understood
     in terms of both the changing intensity of sunlight at a
     location over the year, throughout the day, and changes
     in the length of the day period.

14. A full understanding of season changes must include
     knowledge of the path of the sun across the sky and how
     it varies throughout the year. The location of the sun
     at sunrise, local noon, and sunset, can be observed to
     vary on a daily basis.

KEY TERMS AND CONCEPTS

Provide a short definition or description of the following key
words and concepts from the chapter.

<u>Geometry of the Earth in Space</u>

     Milky Way Galaxy

     Solar System

     Plane of the Ecliptic

     Revolution Around the Sun

     Perihelion and Aphelion

     Rotation About the Axis

Axis Location

Parallelism

Geometry of the Solid Earth

Size

Radius

Diameter

Circumference

Shape

Oblate Ellipsoid

Surface Relief

Distribution of Oceans

Basic Direction

North, South, East, and West

Magnetic North and South

Grid System

Latitude

Parallels

Northern and Southern Hemispheres

Longitude

Meridians

Prime Meridian

Eastern and Western Hemispheres

Degrees, Minutes, and Seconds

Time

    Local Time

        Solar Noon

    Standard Time

        Meridional Framework of 15^0

        Political Changes

        Daylight Saving Time

    International Date Line

Location

    North Pole

        Polar Latitudes

    Arctic Circle

        Middle Latitudes

    Tropic of Cancer

        Tropical Latitudes

    Equator

        Tropical Latitudes

    Tropic of Capricorn

        Middle Latitudes

    Antarctic Circle

        Polar Latitudes

    South Pole

Formation of the Universe

    The "Big Bang" Theory

    Solar Wind

Seasons

Annual Changes in the Earth-to-Sun Distance (Minor Factor)

Annual Changes in the Altitude of Sun (Axis Tilt)

Annual Changes in the Length of Day (Rotation)

Position (Latitude) on Earth

Vernal Equinox

Summer Solstice

Autumnal Equinox

Winter Solstice

Path of the Sun in the Sky

TESTING: MATCHING I

Match the term on the left with the most appropriate phrase or
term on the right.

1.	_____ galaxy		A.	covered by oceans
2.	_____ axis		B.	earth revolution
3.	_____ aphelion		C.	minor surface changes
4.	_____ ecliptic		D.	12,700 kilometers
5.	_____ diameter		E.	line through poles
6.	_____ oblate		F.	assemblage of stars
7.	_____ rotation		G.	farthest away
8.	_____ relief		H.	not round but
9.	_____ circumference		I.	40,000 kilometers
10.	_____ 71%		J.	causes day and night

TESTING: MATCHING II

11.	_____ north or south		A.	magnetic direction
12.	_____ latitude		B.	Greenwich
13.	_____ maximum sun altitude		C.	local noon
14.	_____ one hour jump		D.	poleward
15.	_____ compass north		E.	23 degrees south
16.	_____ prime meridian		F.	parallels
17.	_____ Australia		G.	zero degrees latitude
18.	_____ Capricorn		H.	southern hemisphere
19.	_____ North Pole		I.	axis cuts through surface
20.	_____ Equator		J.	daylight saving time

TESTING: MULTIPLE CHOICE

Choose the best response for each of the following multiple
choice questions.   Each question has only one correct answer.

21.   The earth is closest to the sun on:
      a.   January 3
      b.   March 21
      c.   June 21
      d.   September 21

22.   A location in the Southern hemisphere which experiences
      24 hours of darkness at least one day per year:
      a.   lies in the middle latitudes
      b.   lies north of the Antarctic Circle
      c.   lies between 66 degrees south and the South Pole
      d.   lies equatorward of the Tropic of Capricorn

23.   As one moves from west to east over the International
      Date Line:
      a.   one day is added to the date
      b.   one day is subtracted from the date
      c.   both the date and time of day change
      d.   only the time of day changes

24.   The oblate shape of the geoid is most directly the result
      of the earth:
      a.   revolving about the sun
      b.   rotating about its axis
      c.   the magnetic field of the core
      d.   the ellipticity of the orbit

25.   The length of daylight and night are equal for any spot
      on the globe:
      a.   on the winter solstice of the northern hemisphere
      b.   on the winter solstice of the southern hemisphere
      c.   on either equinox
      d.   the day and night periods never have equal length

26.   A primary cause of the seasons is the tilt of the axis of
      rotation of the earth.   The axis tilt is approximately:
      a. $23 1/2°$ from the vertical
      b. $23 1/2°$ from the horizontal
      c. $23 1/2°$ from a perpendicular to the plane of the
           ecliptic
      d.   parallel to the plane of the ecliptic

27.   The low latitudes are an area of the earth:
      a.   between the Tropics
      b.   poleward of the Arctic and Antarctic Circles
      c.   the middle latitudes
      d.   areas close to the Greenwich meridian

28. The primary cause of seasons is:
     a. ellipticity of the earth's orbit around the sun
     b. the inclination of the axis of rotation of the earth
         from the vertical
     c. the slight tipping of the plane of the ecliptic
     d. altitude on the earth's surface

29. Lines of the grid system of the earth which converge at
     the poles are called _____ of longitude.
     a. parallels
     b. minutes
     c. meridians
     d. great circles

30. The approximate distance covered in moving exactly along
     one degree of latitude is:
     a. 69 miles
     b. 52 miles
     c. 36 miles
     d. 24 miles

31. The earth has a circumference of approximately 24,900
     miles and it rotates once every 24 hours. The speed of
     rotation around the axis is therefore:
     a. 1510 miles per hour
     b. 1040 miles per hour
     c. 720 miles per hour
     d. variable between the equator and the poles

32. At a particular place the altitude of the sun would be
     greatest at:
     a. 12 o'clock noon daylight saving time
     b. 12 o'clock noon standard time
     c. 12 o'clock noon Greenwich time
     d. 12 o'clock noon local time

33. A location for which the length of the day never varies
     from 12 hours is:
     a. found at the Arctic Circle
     b. found at the Tropic of Capricorn
     c. found at the equator
     d. there are no such locations; because of the tilt of
         the axis, day length varies everywhere

34. By following the north arrow of a compass:
     a. one can move towards the north magnetic pole
     b. one can move towards the north pole
     c. one can move along a meridian of longitude
     d. one can move parallel to a line of latitude

35.   One of the most commonly espoused creation theories is:
       a.   the Big Bang theory
       b.   the finite universe theory
       c.   the primordial egg theory
       d.   the supernova theory

TESTING: SHORT ANSWER AND ESSAY

1.   List the factors which result in seasonal changes over the earth.   Which is the most important factor?

2.   Describe the planetary distribution of sunlight on the summer solstice of the northern hemisphere.   Where is the radiation most intense at local noon?   Where is the day period the longest?   At what latitude is the total amount of sunlight greatest?

3.   Why is the spacing between meridians variable, while the spacing of lines of latitude is always 69 miles between each degree?

4. General confusion often surrounds the purpose of the
    International Date Line. Explain the purpose of the date
    line and what happens as one crosses it.

5. The earth revolves around the sun in an orbit that is
    basically circular. How important is the ellipticity of
    this orbit in explaining seasonal climatic variability?
    Why does the northern hemisphere have winter during the
    time when the earth is closest to the sun?

6. Where might you go so that you would have no shadow at 12
    o'clock local noon? Explain your answer and give
    locations after consulting an atlas.

PLACE AND PHYSICAL FEATURE LOCATION

UNITED STATES MAP EXERCISE

Locate the following by name or number on the map of the **United States** (NOTE: you may wish to sketch in the major landforms).

PHYSICAL FEATURES

 1. Central Valley of California
 2. Lake Erie
 3. Sierra Nevada
 4. Rio Grande
 5. Cape Canaveral, Florida
 6. Lake Superior
 7. Lake Ontario
 8. Rocky Mountains
 9. Columbia River
10. Mississippi River
11. Adirondack Mountains
12. Florida Strait
13. Lake Huron
14. Appalachian Mountains
15. Colorado River
16. Lake Michigan
17. Great Salt Lake
18. Ohio River
19. Missouri River
20. Ozark Plateau
21. Great Plains
22. Colorado Plateau
23. Coast Ranges
24. Great Basin
25. Cascade Mountains
26. Oahu
27. Maui
28. Hawaii
29. Alaska Range
30. Brooks Range

MAJOR CITIES

31. New York City
32. Chicago
33. Los Angeles
34. Seattle
35. San Diego
36. Boston
37. Washington, D.C.
38. Miami
39. Portland, Oregon
40. New Orleans
41. St. Louis
42. Honolulu
43. Anchorage
44. Fairbanks
45. Juneau
46. Houston
47. Denver
48. Dallas
49. Atlanta
50. Philadelphia
51. Cincinnati
52. Pittsburgh
53. San Francisco

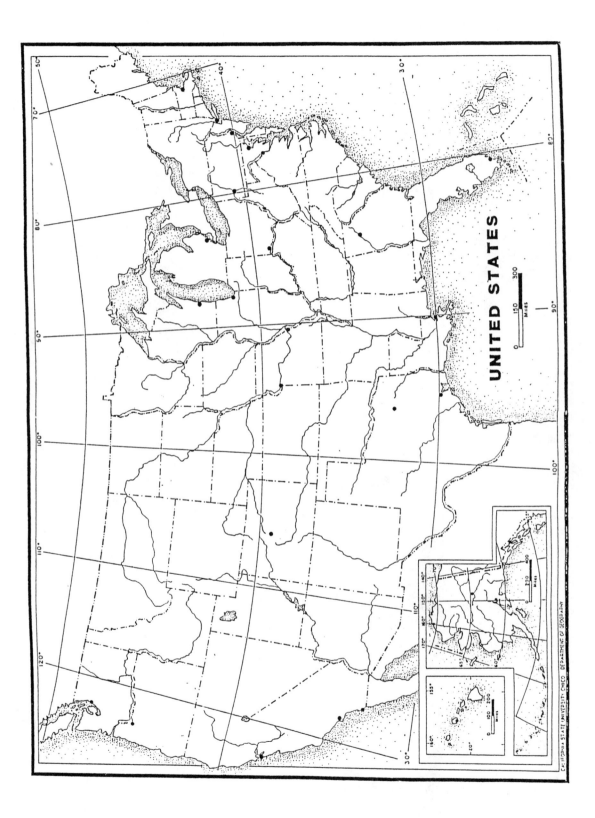

UNITED STATES

C H A P T E R   3

DEPICTING THE EARTH– MAPS AND REMOTE SENSING

LEARNING OBJECTIVES

1. Understand the basic functions of maps and also the basic types of information displayed on maps.

2. Become conversant with the basic map properties, distortions, and scales used to depict the earth's surface on the map surface.

3. Identify the main types of map projections and the important characteristics of each. These would allow a map user to choose a particular projection for a specific purpose.

4. Be able to correct erroneous assumptions that can be gained from displaying the earth on highly distorted projections such as the popular Mercator projection.

5. Describe the main functions and uses of topographic maps that illustrate the relief of the earth's surface.

6. Gain a basic introductory understanding of the use of remote sensing in gathering data for mapping.

7. Acquire an understanding of the use of satellite imagery in gathering data for mapping earth systems, and in the uses of computers for computer mapping of this acquired data.

KEY TERMS AND CONCEPTS

Map

Cartography

Basic Map Information

Title

Date

Location

Directional Orientation

Legend

Map Scale

Verbal Scale

Graphic Scale

Representative Fraction Scale

Small-Scale Map

Large-Scale Map

Map Projection

Standard Points (Lines)

Polar Projections

Conic Projections

Cylindrical Projections

Elliptical Projections

Interrupted Projections

Condensed Projections

Mercator Projections

Planimetric Map

Spot Elevation

Topographic Map

Hachures

Color

Shading

Contour Lines

Raised Relief

Contour Interval

Bench Mark

Index Contours

Remote Sensing

Computer Mapping

Satellites

TIROS

GOES

LANDSAT

Geographic Information Systems

TESTING: TRUE/FALSE

_____   1.   The art and science of map-making is called Remote Sensing.

_____   2.   The Earth's curved surface may be conveniently flattened onto a map without distortion.

_____   3.   A globe is the best way to map the Earth without distortion.

_____   4.   The map scale is the ratio between distances on a map and the corresponding distances on the Earth's surface.

_____   5.   The Mercator projection is a cylindrical map projection.

_____ 6. A conic projection is best suited for depicting middle latitude regions such as the United States with minimum distortion.

_____ 7. Contour lines are used on topographic maps to show horizontal scale.

_____ 8. All topographic maps have the same contour interval.

_____ 9. The deployment or orbiting satellites for data gathering began before World War II.

_____ 10. Geographic information systems can be considered automated atlases of mapped data.

TESTING: MULTIPLE CHOICE

11. A properly prepared map does <u>not</u> need to have:
    a. a title
    b. directional orientation
    c. a scale
    d. colored tints

12. A map scale of 1:63,360 is interpreted as:
    a. one inch equals one mile
    b. 63,360 inches equals one-half mile
    c. one inch equals 63,360 feet
    d. none of the above

13. A good map projection to accurately depict Europe would be a(n):
    a. elliptical projection
    b. cylindrical projection
    c. interrupted projection
    d. conic projection

14. The Mercator projection:
    a. is a commonly used elementary school world map
    b. excessively distorts the high latitudes
    c. shows more of the northern than the southern hemisphere
    d. all of the above

15. Topographic maps:
    a. always have raised relief
    b. use contour lines to express relief
    c. are drawn with hachures
    d. never have colors

16. Contour lines:
     a. are always spaced close together
     b. are always spaced far apart
     c. do not meet or cross one another
     d. are always closed

17. Contour lines point uphill when crossing:
     a. streams
     b. hills
     c. mountains
     d. all of the above

18. The contour interval on a topographic map:
     a. is the horizontal distance between contour lines
     b. is the vertical distance between contour lines
     c. is always 20 feet
     d. is always 50 feet

19. Topographic maps are prepared by and purchased from the:
     a. CIA
     b. DEA
     c. USGS
     d. USDA

20. Which of the following is not a weather satellite:
     a. LANDSAT
     b. GOES
     c. TIROS
     d. all are weather satellites

21. Remote sensing devices include:
     a. cameras
     b. radar
     c. orbiting satellites
     d. all of the above

22. Geographic information systems:
     a. have existed since 1900
     b. are automated atlases of mapped data
     c. only deal with weather data
     d. do not depend upon computers

23. Which of the following is not a common method of
     expressing map scale:
     a. representative fraction
     b. symbol scale
     c. graphic scale
     d. verbal scale

24. Large-scale maps:
     a. show more detail than small scale maps
     b. cover entire continents
     c. show less detail than small scale maps
     d. ar always 1:63,360

25. Computer mapping:
     a. is an aid to modern cartography
     b. is a close technology to remote sensing
     c. can make topographic maps from aerial photographs
     d. all of the above

TESTING: SHORT ANSWER AND ESSAY

1. List and briefly discuss the basic types of map
    information that should be included on any useful map.

2. What is a map projection? How are map projections
    devised? Why should an appropriate map projection be
    chosen to illustrate a particular part of the Earth?

3.   Discuss the use of the Mercator projection from the
     following standpoints: What type of projection is it, the
     types of misinformation carried by it, and, the reasons
     for its popularity. Have alternative map projections been
     devised?

4.   What information is provided by topographic maps and how
     are they used? What are contour lines and what are their
     main characteristics?

5.   What is Remote Sensing? How are remote sensing
     techniques and technology helpful in map-making?

6.   Relate the following to one another: LANDSAT, Geographic
     Information Systems, and Computer Mapping. Does this
     interrelationship give evidence about the manner of
     gathering and storing geographic data, and displaying it
     on modern maps?

PLACE AND PHYSICAL FEATURE LOCATION

CANADA MAP EXERCISE

Locate the following by name or number on the map of **Canada** (NOTE: you may wish to sketch in the major landforms).

PROVINCES

1. Alberta
2. New Brunswick
3. Ontario
4. British Columbia
5. Saskatchewan
6. Newfoundland
7. Northwest Territories
8. Manitoba
9. Yukon Territory
10. Nova Scotia
11. Quebec
12. Prince Edward Island

MAJOR CITIES

33. Ottawa
34. Montreal
35. Vancouver
36. Victoria
37. Toronto
38. Edmonton
39. Quebec
40. Winnipeg

PHYSICAL FEATURES

13. Coast Ranges
14. James Bay
15. Lake Superior
16. Strait of Juan de Fuca
17. Hudson Strait
18. Baffin Island
19. Fraser River
20. Lake Ontario
21. Rocky Mountains
22. Gulf of St. Lawrence
23. Mackenzie River
24. Lake Huron
25. St. Lawrence River
26. Vancouver Island
27. Hudson Bay
28. Yukon River
29. Lake Erie
30. Lake Winnipeg
31. Great Bear Lake
32. Great Slave Lake

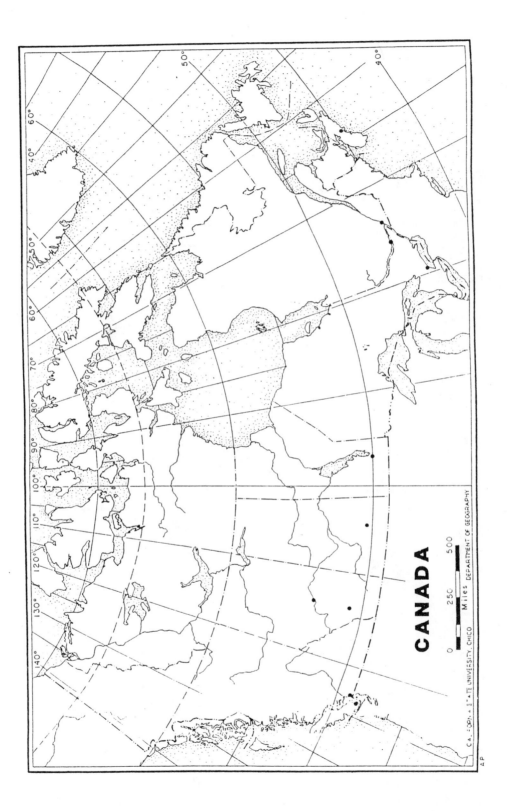

CANADA

0   250   500
Miles   DEPARTMENT OF GEOGRAPHY

CALIFORNIA STATE UNIVERSITY, CHICO

C H A P T E R   4

EARTH'S ATMOSPHERIC ENVELOPE

LEARNING OBJECTIVES

1. Explain the role played by the earth's atmosphere in protecting the earth's surface from meteorites and cosmic radiation.

2. Describe a theoretical origin of the earth's atmosphere as it might interrelate with volcanic gases.

3. Identify the most abundant <u>nonvariable</u> and <u>variable gases</u> in the earth's atmosphere and their relative abundance.

4. Outline the steps in the cycling of such atmospheric gases as carbon dioxide, oxygen, and water vapor.

5. Discuss the <u>ozone layer</u> of the earth's atmosphere in terms of: the nature and source of ozone, the location of the ozone layer, the effect of ozone on incoming solar radiation, and human impacts that can act to deplete the ozone layer.

6. Explain the importance of particulate matter in the earth's atmosphere as <u>condensation nuclei</u>.

7. Differentiate between the four major thermal layers of the atmosphere and the reasons for their existence.

8. Differentiate between <u>weather</u> and <u>climate</u> as these refer to atmospheric phenomena.

9.   Note the importance of the <u>troposphere</u> as the most important layer of the atmosphere for weather systems.

10. Compare and contrast urban air pollution with agricultural air pollution in terms of atmospheric changes.

11. Note the interrelationships between air pollution patterns and atmospheric <u>temperature inversions</u>.

12. Discuss the uniqueness of the earth's atmosphere when compared with the types of atmospheres existing on the other eight planets of our solar system.

KEY TERMS AND CONCEPTS

Cosmic Radiation

Nonvariable Gases

Variable Gases

Ozone (Ozone Layer)

Particulates

Condensation Nuclei

Homosphere

Heterosphere

Troposphere

Tropopause

Stratosphere

Stratopause

Mesosphere

Mesopause

Thermosphere

Weather (Meteorology)

Climate (Climatology)

Temperature Inversion

TESTING: MATCHING

A.   OXYGEN

B.   CARBON DIOXIDE

C.   STRATOSPHERE

D.   NITROGEN

E.   DESERTIFICATION

F.   TROPOSPHERE

G.   CLIMATOLOGY

H.   COSMIC RADIATION

I.   VOLCANIC DUST

J.   MESOSPHERE

K.   OZONE

L.   METEOROLOGY

M.   TROPOPAUSE

N.   TEMPERATURE
        INVERSION

O.   VARIABLE GASES

_____ 1.   A source of condensation nuclei in the earth's atmosphere

_____ 2.   The layer at the top of the troposphere

_____ 3.   Produces ozone by striking oxygen molecules

_____ 4.   The atmospheric layer which contains most weather processes

_____ 5.   An atmospheric gas which can absorb heat

_____ 6.   An abundant but inert gas in the earth's atmosphere

_____ 7.   A reactive gas present to the extent of 21% in the atmosphere

_____ 8.   Occurs when warm air overlies cold air

_____ 9.   Triatomic oxygen

_____ 10. The study of weather

_____ 11. The atmospheric layer which contains the ozone layer

_____ 12. A process caused by vegetation and soil removal

TESTING: MULTIPLE CHOICE

13. Present-day volcanos release gases in these proportions:
     a.   50% oxygen and 50% nitrogen
     b.   85% water vapor, 10% carbon dioxide and 1-2% nitrogen
     c.   75% carbon dioxide and 25% nitrogen
     d.   50% carbon dioxide and 50% oxygen

14. The three main nonvariable gases which comprise over 98%
     of the earth's atmosphere are:
     a.   carbon dioxide, oxygen, and helium
     b.   nitrogen, carbon dioxide and water vapor
     c.   oxygen, ozone and carbon dioxide
     d.   nitrogen, oxygen and argon

15. In the earth's atmosphere oxygen is present to the extent
     of:
     a.   21%
     b.   50%
     c.   95%
     d.   0.093%

16. Both of the following gases are inert:
     a.   oxygen and carbon dioxide
     b.   nitrogen and argon
     c.   oxygen and argon
     d.   carbon dioxide and ozone

17. Oxygen:
     a.   is liberated by volcanic eruptions
     b.   enters into chemical reactions with rock materials
     c.   is unrelated to ozone
     d.   is the most abundant gas in the earth's atmosphere

18. Carbon dioxide:
     a.   is essential to plants in their photosynthetic
          processes
     b.   is decreasing in abundance in the earth's atmosphere
     c.   is present to the extent of 34% in the earth's
          atmosphere
     d.   is not involved in a cycle of interchange with oxygen

19. Nitrogen:
     a.   is a highly reactive gas
     b.   is not produced by volcanic eruptions
     c.   comprises 98% of the earth's atmosphere by weight
     d.   is the most abundant gas in the earth's atmosphere

20. Water vapor:
    a. is water in the liquid state
    b. is found most abundantly near the earth's surface
    c. varies independently with air temperature
    d. is more abundant in cold rather than warm air

21. Which of the following does <u>not</u> refer to ozone:
    a. ozone is triatomic oxygen
    b. most ozone exists in a layer between altitudes of 10-
        40 miles
    c. it is produced by the bombardment of oxygen by cosmic
        radiation
    d. once produced, ozone is transparent to incoming
        shortwave length radiation

22. CFC's in the earth's atmosphere:
    a. Are created by human industrial processes.
    b. Are relatively stable gases with chemical half-lives
        greater than 75 years.
    c. Can react with sunlight in the stratosphere to
        decrease ozone.
    d. All of the above.

23. The troposphere:
    a. is the highest thermal layer of the atmosphere
    b. lies above the tropopause
    c. contains virtually no clouds
    d. is the zone where most observable weather phenomena
        occur

24. The stratosphere:
    a. lies between the troposphere and the mesosphere
    b. lies below the tropopause
    c. is above the zone layer
    d. is a layer of cooling in the earth's atmosphere

25. The mesosphere:
    a. lies between the tropopause and the stratopause
    b. contains the atmosphere's "ozone layer"
    c. contains the atmosphere's highest temperature values
    d. is the third of the atmosphere's thermal layers

26. The thermosphere:
    a. is an outer zone of very high temperatures
    b. is the primary zone of weather phenomena
    c. is below 50 miles in the earth's atmosphere
    d. lies within the "homosphere"

27. A temperature inversion:
    a. is unrelated to heating and cooling of the earth's
        surface
    b. occurs when cold air is trapped atop warm air
    c. is a reversal of the normal tropospheric pattern of
        declining temperatures with elevation
    d. only occurs in the stratosphere

28. With regard to the atmospheres of the other planets in
    our solar system:
    a. only the earth has an atmosphere
    b. other than Earth, only Mars has sufficient
        atmospheric oxygen to sustain life
    c. Jupiter's atmosphere is comprised of mainly hydrogen
        and helium
    d. Mercury and Venus have identical atmospheres

29. Which of the following does not refer to the homosphere
    of the earth's atmosphere:
    a. it contains over 99.9% of the total air
    b. it has a relatively constant gaseous composition
    c. there is no mixing by vertical wind currents
    d. it comprises the lowermost 50 miles of the
        atmosphere

30. Urban air pollution is affected by:
    a. the burning of coal
    b. atmospheric inversions
    c. foggy weather
    d. automobile use
    e. all of the above

1. In what ways does the atmosphere interact with processes
    at the earth's surface?

2. List and briefly identify the <u>five</u> main sources of parti-
    culate matter in the earth's atmosphere. What is the
    importance of this material?

3. Why does the thickness of the troposphere vary with
    latitude? What atmospheric processes are occurring to
    cause this variation? Compare the thickness of the
    equatorial troposphere to the polar troposphere?

4. In what ways are humans affecting the ozone layer? What
    lines of evidence indicates that such alternatives are
    taking place? What are the major consequences of these
    impacts?

5. Discriminate between weather and climate. Although both
    would be studying the earth's atmospheric processes, in
    what major ways would a meteorologist be taking a
    different approach than a climatologist?

6. What are the main processes which occur in urban areas to
    cause atmospheric pollution? Are there any human health
    hazards which are involved?

7. What types of processes are implicated in the topic of
    "agricultural air pollution"? How can these be related
    to the process of <u>desertification?</u>

8. Discuss the changing thermal processes which occur as one
    moves from the earth's surface through the four major
    thermal layers of the earth's atmosphere. How do these
    changes relate to both gravity and incoming solar
    radiation?

PLACE AND PHYSICAL FEATURE LOCATION

CANADA MAP EXERCISE

Locate the following by name or number on the map of **Canada**
(NOTE: you may wish to sketch in the major landforms).

PROVINCES

1.   Alberta
2.   New Brunswick
3.   Ontario
4.   British Columbia
5.   Saskatchewan
6.   Newfoundland
7.   Northwest Territories
8.   Manitoba
9.   Yukon Territory
10. Nova Scotia
11. Quebec
12. Prince Edward Island

MAJOR CITIES

33. Ottawa
34. Montreal
35. Vancouver
36. Victoria
37. Toronto
38. Edmonton
39. Quebec
40. Winnipeg

PHYSICAL FEATURES

13. Coast Ranges
14. James Bay
15. Lake Superior
16. Strait of Juan de Fuca
17. Hudson Strait
18. Baffin Island
19. Fraser River
20. Lake Ontario
21. Rocky Mountains
22. Gulf of St. Lawrence
23. Mackenzie River
24. Lake Huron
25. St. Lawrence River
26. Vancouver Island
27. Hudson Bay
28. Yukon River
29. Lake Erie
30. Lake Winnipeg
31. Great Bear Lake
32. Great Slave Lake

MIDDLE AMERICA

MILES

0   250   500   750   1000

CALIFORNIA STATE UNIVERSITY, CHICO • DEPARTMENT OF GEOGRAPHY

C H A P T E R   5

ENERGY FLOW AND AIR
TEMPERATURE

1. Define and describe the three major energy transfer
    processes of radiation, conduction, and convection.

2. Describe what occurs to solar radiation as it passes
    through the earth's atmosphere.

3. Identify the variable mechanisms whereby insolation is
    either absorbed or reflected (Albedo) by the earth's
    surface.

4. Discuss the conversion of solar radiation to terrestrial
    radiation and the significant differences between them.

5. Explain the interrelationship between terrestrial
    radiation and the atmosphere in providing the "greenhouse
    effect".

6. Describe the relationship between mean sun altitudes and
    the long-term global temperature pattern.

7. Identify the key temperature statistics that can be
    calculated from temperature data and that can be used to
    provide climatic information for a place.

8. Identify and discuss the five primary controlling factors
    that influence temperature over the earth.

9. Compare and contrast the specific heat of water with that
    of land and explain the consequent temperature
    differences between the two substances.

10. Differentiate between continental and maritime climatic characteristics.

11. Describe the temperature and climatic impacts of ocean currents, both warm and cold, upon their respective coasts.

12. Discuss the impacts of topography and slope characteristics upon temperature patterns.

13. Know the direct general human impacts upon world temperature patterns and specifically the relationship between an increase in atmospheric carbon dioxide and the greenhouse effect.

14. Identify the interrelationships between alterations in global vegetation patterns and resultant surface energy.

15. Describe the characteristics of electromagnetic radiation and its relationship to the temperature of a radiating body.

16. Differentiate between the three temperature scales in common use -- Fahrenheit, Celsius, and absolute; and, be able to convert temperature values between them.

KEY TERMS AND CONCEPTS

Temperature

Kinetic Energy

Potential Energy

Radiation

Conduction

Convection

Electromagnetic Spectrum

Insolation

Albedo

Latent Heat

Terrestrial Radiation

Greenhouse Effect

Energy Balance

Isotherm

Daily Mean Temperature

Monthly Mean Temperature

Daily Temperature Range

Annual Temperature Range

Emissivity

Specific Heat

Average Environmental Lapse Rate

Leeward Coasts

Windward Coasts

Fahrenheit Scale

Celsius Scale

Kelvin (Absolute) Scale

TESTING: MATCHING

A.   RADIATION   _____   1.   Solar radiation

B.   ALBEDO   _____   2.   Water freezes at 32 degrees

C.   ISOTHERM   _____   3.   Has a predominantly water-
                                         to-land wind flow

D.   CONDUCTION

E.   GREENHOUSE EFFECT   _____   4.   Heat transfer with wind
                                         movement vertically

F.   ADVECTION   _____   5.   Due to the absorption of
                                         carbon dioxide

G.   FAHRENHEIT SCALE

H.   CONVECTION   _____   6.   Electromagnetic wave trans-
                                         mission of energy

I.   INSOLATION   _____   7.   Connects points with the
                                         same temperature

J.   TERRESTRIAL
         RADIATION   _____   8.   Energy reflected by a
                                         surface

K.   EMISSIVITY
                          _____   9.   Water boils at 100 degrees

L.   SPECIFIC HEAT
                          _____   10.   Longwave length radiation

M.   LEEWARD COAST
                          _____   11.   Occurs over the cold
N.   CELSIUS SCALE   California current

O.   PERIHELION   _____   12.   Heat movement from one
                                         substance to another

TESTING: MULTIPLE CHOICE

13.   The process of energy transmission by means of electro-
      magnetic waves is:
      a.   conduction
      b.   convection
      c.   radiation
      d.   advection

14.   Energy transfer through physical movement of air is
      termed:
      a.   advection
      b.   radiation
      c.   conduction
      d.   convection

15. Heat transfer by physical contact between land and air
    is:
    a. conduction
    b. advection
    c. convection
    d. radiation

16. Solar energy is:
    a. longwave radiation
    b. shortwave radiation
    c. different from insolation
    d. heavily altered before contact with the earth's
        atmosphere

17. The earth's atmosphere affects incoming solar radiation
    by means of:
    a. redirecting 99.97% back to space
    b. almost total absorption by clouds
    c. absorption only with no appreciable albedo of
        atmospheric constituents
    d. scattering by air molecules and dust particles

18. The albedo of the earth system as a whole is:
    a. 99.97%
    b. 78%
    c. 30%
    d. less than 1%

19. The following surface would have the lowest albedo:
    a. concrete
    b. fresh snow
    c. sand
    d. a dark, moist soil

20. Terrestrial radiation is:
    a. longwave radiation
    b. shortwave radiation
    c. insolation
    d. all energy reflected by the earth's surface

21. The two atmospheric gases which most efficiently absorb
    terrestrial radiation are:
    a. carbon monoxide and ozone
    b. nitrogen and argon
    c. oxygen and carbon monoxide
    d. carbon dioxide and water vapor

22. The "greenhouse effect":
    a. results in warming of the lower atmosphere
    b. cannot be altered by human activities
    c. relates to dust in the earth's atmosphere
    d. occurs primarily in the "ozone layer"

23. Lines connecting places with the same temperature are called:
    a. isobars
    b. isohyets
    c. isotherms
    d. thermograms

24. The daily mean temperature is the:
    a. difference between the daily high and low temperatures
    b. the highest temperature during a 24-hour period
    c. addition of the daily high and low temperatures divided by two
    d. average of hourly temperatures

25. The most important factor controlling air temperatures at the earth's surface is:
    a. the "greenhouse effect"
    b. the altitude of the sun
    c. the elevation of a place
    d. the albedo

26. The specific heat of water:
    a. is higher than land
    b. is lower than land
    c. is proportional to the albedo
    d. is usually equal to land

27. Compared with water, land:
    a. loses more heat by evaporation
    b. is less opaque to solar radiation
    c. is more mobile and heats to a great depth
    d. tends to have larger daily and annual temperature ranges

28. Compared with maritime locations, continental areas:
    a. have warmer winters
    b. have greater annual temperature ranges
    c. have cooler summers
    d. have smaller annual temperature ranges

29. The average environmental lapse rate is:
    a. the rate of cooling as one ascends a mountain slope
    b. the general cooling rate at high elevations
    c. a tropospheric temperature decrease of 3.5°F per 1,000 feet
    d. a stratospheric temperature increase of 3.5°F per 1,000 feet

30. Windward coasts:
    a. do not usually experience maritime air masses
    b. are affected offshore air masses
    c. have a dominant onshore windflow
    d. have land to water windflow

31. The British Isles and Iceland have generally warmer temperatures due to:
    a. their low elevations
    b. the North Atlantic Drift
    c. the cold ocean currents of the North Atlantic Ocean
    d. their subtropical locations

32. Surface temperatures are lower when:
    a. the sun is low in the sky
    b. no dust is in the atmosphere
    c. no cloud cover exists to block insolation
    d. a mountain slope faces towards the sun

TESTING: SHORT ANSWER AND ESSAY

1.   What are the relationships between mean sun altitudes and the long-term global temperature patterns? Are there any short-term patterns which are also controlled by the sun's altitude?

2.   What processes occur as solar radiation passes through the earth's atmosphere and arrives at the surface?

3.   What is terrestrial radiation? How is it important in describing the temperature patterns of the earth's surface and the lower atmosphere?

4. Compare and contrast the specific heat of land with that
    of water. What are the climatic impacts of this
    difference in terms of a comparison between maritime and
    mid-continental localities?

5. Which sides of continents are most commonly paralleled by
    cold currents? By warm currents? Why are windward
    coasts more strongly influenced by ocean currents than
    are leeward coasts?

6. List and briefly explain the five major primary controls
    on the earth's temperature patterns. Are there any
    secondary temperature influences that might be
    considered?

7. What is the value of the average environmental lapse
    rate, and where does this occur? When should this lapse
    rate be taken under consideration?

8. What human activities could cause alterations in the
    earth's temperature patterns? How do these relate to
    naturally occurring processes either in the atmosphere
    or at the earth's surface?

9. How does solar radiation differ from terrestrial
    radiation? Why is there a difference?

10.   List the major ways that cloud cover affects temperature
      patterns over the earth.

11.   List the major topographic influences on temperatures.
      How does slope and the slope orientation to air flow
      affect temperature?

PLACE AND PHYSICAL FEATURE LOCATION

SOUTH AMERICA MAP EXERCISE

Locate the following by name or number on the map of **South America** (NOTE: you may wish to sketch in the major landforms).

COUNTRIES MAJOR CITIES

1.   Venezuela
2.   Chile
3.   Ecuador
4.   Uruguay
5.   Peru
6.   Brazil
7.   Guyana
8.   Bolivia
9.   French Guiana
10. Colombia
11. Argentina
12. Surinam
13. Paraguay
14. Trinidad and Tobago

35. Montevideo
36. Sao Paulo
37. Rio de Janeiro
38. Lima
39. Recife
40. La Paz
41. Asuncion
42. Bogota
43. Buenos Aires
44. Caracas
45. Guayaquil
46. Brasilia, D.F.
47. Santiago
48. Quito

PHYSICAL FEATURES

15. North Atlantic Ocean
16. Caribbean Sea
17. Andes Mountains
18. South Atlantic Ocean
19. Gulf of Guayaquil
20. Brazilian Highlands
21. Cape Horn
22. Amazon River
23. Pacific Ocean
24. Falkland Islands
25. Lake Maracaibo
26. Orinoco River
27. Rio de la Plata
28. Pampas
29. Altiplano (Bolivia)
30. Guyana Highlands
31. Patagonian Plateau
32. Llanos
33. Galapagos Islands
34. Lake Titicaca

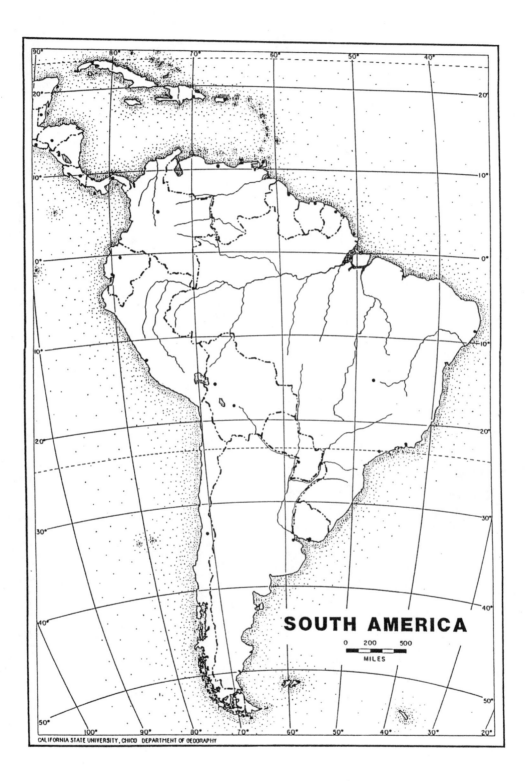

SOUTH AMERICA

0   200   500

MILES

CALIFORNIA STATE UNIVERSITY, CHICO DEPARTMENT OF GEOGRAPHY

AIR PRESSURE AND WIND

LEARNING OBJECTIVES

1. Develop a familiarity with the basic vocabulary dealing with air pressure and wind systems.

2. Investigate the meaning of air pressure and the factors which cause air pressure to rise and fall over time.

3. Devise basic rules which relate pressure and pressure changes to the daily weather.

4. Visualize the portrayal of basic weather data on maps and understand how weather predictions can be made from these maps.

5. Study the characteristics of high and low pressure cells and relate these characteristics to weather changes.

6. Generalize daily weather into global patterns of air pressure for January and for July; relate these pressure belts to the distribution of climates around the world.

7. Attempt to develop an intuitive understanding of the Coriolis effect and its effect on wind direction.

8. Consider the effect of friction with the ground surface on the direction and the speed of winds.

9. Become sufficiently familiar with forces producing winds (pressure gradient, Coriolis effect, and friction) to predict resultant wind direction.

10. Synthesize generalizations about the production of winds and the movement of air masses in a three-dimensional visualization of pressure and wind belts on a rotating earth.

11. Visualize and map the general wind circulation of the earth's atmosphere.

12. Consider the surface wind belts of the globe at key latitudes, between 5°N and 5°S, between 30°-35° N and S, at 60°-65° N and S, and near both poles.

13. Develop an understanding of the major regional and local wind systems.

14. Develop an appreciation of the major upper-level winds of the earth's atmosphere and how these jet streams might influence surface weather conditions.

15. Develop an understanding of why weather prediction is so complex.

KEY TERMS AND CONCEPTS

Provide a short definition or description of the following key words and concepts from the chapter.

Air Pressure

Isobars

Millibars

Anticyclone

Cyclone

Pressure Gradient

Thermal High Pressure

Thermal Low Pressure

Wind Speed and Direction

  Coriolis Effect

  Pressure Gradient Force

  Friction

  Wind Turbulence

Global Pressure

  Inter-Tropical Convergence Zone (ITCZ)

  Subtropical Highs

  Subpolar Lows

  Polar Highs

Global Wind Belts

  Hadley Cell

  Equatorial Doldrums

  Northeast and southeast Trade Winds

  Horse Latitudes

  Middle Latitude Westerlies

  Polar Cell

  Polar Easterlies

  Polar Front Zone

Local and Regional Winds

  Land Breeze

  Sea Breeze

  Chinook (Foehn)

  Monsoon

    Winter Monsoon

    Summer Monsoon

<u>Upper-Level Winds</u>

    Jet Streams

        Subpolar Jet

        Subtropical Jet

<u>Local Winds</u>

    Sea Breeze

    Land Breeze

    Valley Breeze

    Mountain Breeze

    Chinook

    Foehn

    Santa Ana

    Summer Monsoon

    Winter Monsoon

<u>Wind Measurement</u>

    Wind vane

    Anemometer

    Radiosonde

<u>TESTING: TRUE/FALSE</u>

_____   1. Monsoon refers to a marked change in precipitation.

_____   2. Under the guidance of the pressure gradient force winds would blow from low to high pressure.

_____   3. The least important of the three main forces controlling wind speed and direction is the Coriolis force.

_____   4. A simple term for air in motion is 'wind'.

_____ 5. In general, land bodies heat and cool more rapidly than water bodies of a similar size.

_____ 6. The air pressure at sea level is approximately 1.47 pounds per square inch.

_____ 7. A barometer is an instrument used to measure the current air pressure. A barometer showing dropping pressure indicates that inclement weather is on the way.

_____ 8. The primary factor controlling wind direction is the orientation of the pressure gradient.

_____ 9. The sense of direction of the Coriolis effect is counterclockwise in the Southern Hemisphere.

_____ 10. A sea breeze will generally develop during the morning as the land heats up and the air above it rises.

TESTING: MATCHING

11. _____ trade winds

12. _____ jet stream

13. _____ equatorial low

14. _____ doldrums

15. _____ chinook

16. _____ isobars

17. _____ westerlies

18. _____ thermal high

19. _____ polar front

20. _____ Icelandic low

21. _____ Hadley cell

A. semipermanent low pressure

B. lines on a weather map

C. ITCZ

D. surface cooling

E. equatorial variable winds and calms

F. zone of air mass collision

G. local mountain wind

H. tropical easterlies

I. winds of the middle latitudes

J. upper level wind

K. tropical circulation

TESTING: MULTIPLE CHOICE

Choose the best response for each of the following multiple
choice question. Each question has only one correct answer.

22. The direction of the wind is most controlled by:
     a. the pressure gradient distribution
     b. the Coriolis effect
     c. the frictional component
     d. local topography

23. The speed of the wind is most controlled by:
     a. the pressure gradient distribution
     b. the Coriolis effect
     c. the frictional component
     d. local topography

24. The three forces controlling wind direction and wind
     speed, in order of importance, are:
     a. pressure gradient, Coriolis, friction
     b. pressure gradient, friction, Coriolis
     c. Coriolis, pressure gradient, friction
     d. Coriolis, friction, pressure gradient

25. The force which alters the speed and direction of wind
     blowing close to the ground is:
     a. the pressure gradient
     b. the Coriolis effect
     c. the friction
     d. local topography

26. The circulation of the tropics is dominated by:
     a. strong monsoon winds in each hemisphere
     b. equatorward moving winds blowing from high pressure
         to low pressure
     c. irregular, variable winds of the doldrums
     d. the vigorous winds of the subtropical jet stream

27. In general, high pressure is associated with:
     a. upwards moving air
     b. inclement weather
     c. downwards moving air
     d. horizontal winds of enormous strength

28. The horizontal pressure gradient:
     a. is considerably stronger than the vertical pressure
         gradient
     b. results in horizontal air motion called wind
     c. can be disregarded in discussions of weather
     d. is at least 10 bars per kilometer

29. The sense of direction of the Coriolis force in the
    northern hemisphere is:
    a. clockwise
    b. counterclockwise
    c. clockwise during the summer, reversing in the winter
    d. is directed from high pressure to low pressure

30. The westerly wind belt:
    a. is stronger in velocity than the trade winds
    b. is prone to the development of traveling high and low
        pressure systems
    c. is more persistent than the trade winds
    d. has winds that blow from the east to the west

31. Quickly cooling land which fosters the formation of an
    overlying mass of cool and dense air gives rise to:
    a. a monsoon
    b. a chinook
    c. a sea breeze
    d. a land breeze

32. A special type of wind arising from the changes in the
    regional pressure gradient is the:
    a. mountain breeze
    b. monsoon
    c. sea breeze
    d. chinook

33. The critical factor in the designation of a monsoon is
    the:
    a. seasonal reversal of wind direction
    b. sudden onset of a wet season
    c. rapid formation of clouds and precipitation
    d. proximity of a large water body

34. The high velocity core of winds in the upper-level
    westerlies is called the:
    a. jet stream
    b. tropical easterlies
    c. enigmatic gusts
    d. tornados

35. The amount of precipitation associated with the monsoon
    of India is:
    a. 10%
    b. 25%
    c. 40%
    d. 85%

36.   The instrument used to measure wind velocity is a(n):
      a.   barometer
      b.   hygrometer
      c.   thermometer
      d.   anemometer

TESTING: SHORT ANSWER AND ESSAY

1. Weather can be interpreted as a group of phenomena arising from variations in air pressure. What causes these variations to occur?

2. Describe the general circulation of the atmosphere from the equator to the poles? Why does the atmosphere circulate in this way?

3. List and describe the names of the different types of local winds which result from differential heating of land and water. How do such winds resemble such large-scale regional wind systems such as the monsoon systems of Eastern Asia?

4. Why is there a strong relationship between the location
    of semipermanent pressure belts and the global wind
    belts?

5. List everything that you understand about the nature of
    the Coriolis effect? How does the Coriolis effect vary
    in its impact upon winds from low to high latitudes?

6. Where and why do jet streams form? Are they important in
    determining daily weather patterns?

PLACE AND PHYSICAL FEATURE LOCATION

EUROPE MAP EXERCISE

Locate the following by name or number on the map of **Europe**.
(NOTE: you may wish to sketch in the major landforms).

COUNTRIES   **PHYSICAL FEATURES**

1.	Poland	34.	Po River
2.	Finland	35.	Sicily
3.	Norway	36.	Bosporus
4.	Bulgaria	37.	Black Sea
5.	Yugoslavia	38.	Ionian Sea
6.	Cyprus	39.	Elbe River
7.	Italy	40.	Gulf of Finland
8.	Netherlands	41.	The Alps
9.	Malta	42.	Thames River
10.	Albania	43.	Crete
11.	Sweden	44.	Mediterranean Sea
12.	East Germany (Before 1990)	45.	Dardanelles
13.	Switzerland	46.	Iberian Peninsula
14.	Republic of Ireland	47.	Strait of Gibraltar
15.	Spain	48.	Seine River
16.	Luxembourg	49.	Rhine River
17.	England	50.	Baltic Sea
18.	Denmark	51.	Aegean Sea
19.	Czechoslovakia	52.	North Sea
20.	Greece	53.	Adriatic Sea
21.	Austria	54.	English Channel
22.	Vatican City	55.	Danube River
23.	Monaco	56.	Pyrenees
24.	Portugal	57.	British Isles
25.	France	58.	Bay of Biscay
26.	United Kingdom	59.	Corsica
27.	Romania	60.	Sardinia
28.	Iceland	61.	Rhone River
29.	Hungary	62.	Tyrrhenian Sea
30.	Liechtenstein	63.	Carpathian Mts.
31.	Belgium	64.	Jura Mountains
32.	Turkey		
33.	West Germany		

MAJOR CITIES

65.   Paris
66.   Warsaw
67.   Vienna
68.   Reykjavik
69.   Helsinki
70.   Bonn
71.   Budapest
72.   Athens
73.   Rome
74.   Bucharest
75.   Lisbon
76.   Copenhagen
77.   Naples
78.   Bern
79.   London
80.   Ankara
81.   Brussels
82.   Amsterdam
83.   Oslo
84.   Dublin
85.   Berlin
86.   Belgrade
87.   Venice
88.   Milan
89.   Geneva
90.   Munich
91.   Frankfurt

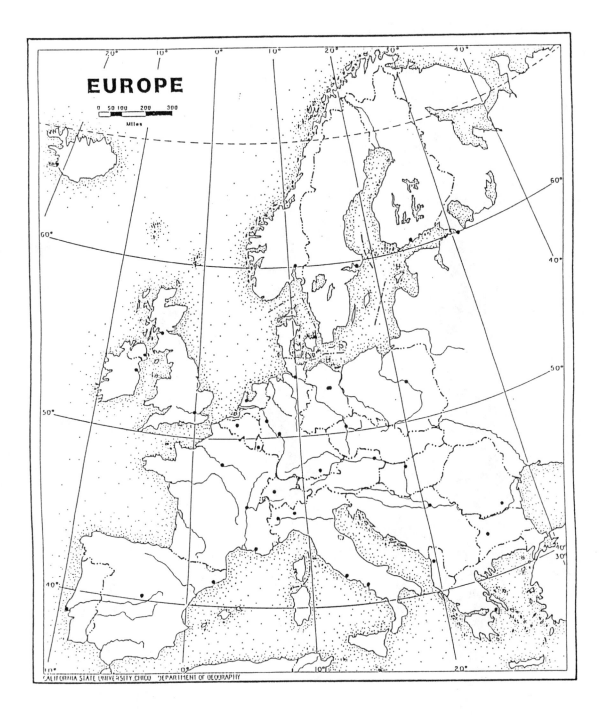

EUROPE

0 50 100   200   300
Miles

C H A P T E R   7

ATMOSPHERIC MOISTURE

1. Describe the absorption or release of heat energy as
    water undergoes the phase changes of evaporation,
    condensation, melting, freezing and sublimation.

2. Explain the five major steps in the hydrologic cycle
    -- the continuous cycle of water exchange between the
    earth's surface and the atmosphere.

3. Outline the basic distribution of water supplies over the
    earth-atmosphere system.

4. Explain the roles of evaporation and condensation in the
    global distribution of energy.

5. Discuss the mechanisms by which water vapor (humidity) is
    held by an air mass and describe the terms which refer to
    this process.

6. Describe the adiabatic changes that occur when air
    ascends of sinks.

7. Explain why understanding the relationship between air
    stability and instability is important in understanding
    cloud development.

8. Compare and contrast the processes which occur to form
    clouds, dew, frost and fog.

9. Differentiate between the environmental circumstances
    which must be present in order to form radiation fog and
    advection fog.

10. Outline the four basic types of cloud formations that are
    based on height of formation in the earth's atmosphere.

11. Relate the four basic cloud formations to air movement
    and resultant weather conditions.

12. Discuss the basic mechanism whereby cloud droplets
    coalesce and precipitation is formed.

13. Compare and contrast the processes which occur in order
    to form rain, sleet, snow and hail.

14. Differentiate between convectional, orographic, frontal
    and cyclonic lifting of air and the conditions which lead
    to each.

15. Describe the basic characteristics and causes of the
    global pattern of precipitation.

16. Explain the importance of the potential evapotranspira-
    tion rate in affecting surface water supplies.

KEY TERMS AND CONCEPTS

Evaporation

Condensation

Sublimation

Heat of Condensation

Hydrologic Cycle

Transpiration

Evapotranspiration

Humidity

Specific Humidity

Relative Humidity

Dew Point

Adiabatic Cooling

Adiabatic Warming

Stable Air

Unstable Air

Dew

Frost

Radiation Fog

Advection Fog

Cloud Types:

    Cumulus

    Stratus

    Cirrus

    Nimbostratus

    Cumulonimbus

Bergeron process

Rain - drizzle and freezing rain

Snow

Sleet

Hail

Convectional Lifting

Orographic Lifting

Frontal Lifting

Cyclonic Lifting

Potential Evapotranspiration Rate

TESTING: MATCHING

(Cloud Types)

A. CUMULUS   _____   1. High, wispy ice crystal clouds

B. NIMBOSTRATUS   _____   2. A cloud type intermediate
                                  between stratus and cumulus
C. CIRRUS
                      _____   3. Small, circular cloud masses
D. ALTOCUMULUS   in a linear pattern

E. STRATUS   _____   4. Associated with long periods
                                  of continuously falling
F. CUMULONIMBUS   precipitation

G. CIRROSTRATUS   _____   5. Forms a halo around the sun

H. STRATOCUMULUS   _____   6. A high layer of gray cloud
                                  covering the entire sky
I. ALTOSTRATUS
                     _____   7. Puffy cloud masses with flat
J. CIRROCUMULUS   bases

                     _____   8. Small, rounded iridescent ice
                                  crystal clouds

                     _____   9. The lowest cloud type often
                                  associated with a fog layer

                     _____   10. Associated with heavy rain,
                                  thunder and hail

TESTING: MATCHING

(Forms of Condensation and Precipitation)

A. FOG

B. CLOUDS

C. FROST

D. DEW

E. RAIN

F. DRIZZLE

G. SNOW

H. HAIL

I. SLEET

J. FREEZING RAIN

_____ 11. Liquid droplets formed on the surface by radiational cooling

_____ 12. Cover 52% of the world at any given time

_____ 13. Particles of ice associated with thunderstorms

_____ 14. Pellets of frozen rain

_____ 15. Formed by sublimation

_____ 16. Crystalline ice in a branded hexagonal form

_____ 17. Formed by both radiation and advection processes

_____ 18. Falling droplets with diameters less than 0.02 inches

_____ 19. Can cause widespread damage to power lines and trees

_____ 20. The most common form of precipitation

TESTING: MULTIPLE CHOICE

21. When the process of evaporation occurs:
   a. energy is released
   b. energy is transferred from the atmosphere to the surface
   c. liquid water changes to water vapor
   d. water vapor changes to liquid water

22. When the process of condensation occurs:
   a. energy is absorbed
   b. energy is transferred from the surface to the atmosphere
   c. liquid water changes to water vapor
   d. water vapor changes to liquid water

23. Sublimation refers to:
     a. the melting of ice
     b. ice changing directly to water vapor
     c. the freezing of water
     d. any process where no energy transfer occurs

24. In the hydrologic cycle:
     a. solar energy is spread more evenly over the earth's
         surface
     b. water is transported only from the continents to the
         oceans
     c. water is transported only from the oceans to the
         continents
     d. only atmospheric water is considered

25. In the distribution of global near-surface water
     supplies:
     a. fresh water supplies comprise over 33%
     b. atmospheric moisture comprises over 73%
     c. rivers and lakes contain about 32%
     d. almost 97% of total water is in the oceans

26. Water evaporated from plants is called:
     a. evapotranspiration
     b. potential evapotranspiration
     c. transpiration
     d. specific Humidity

27. The ratio of the mass of water vapor in a parcel of air
     to the total mass of the air is termed the:
     a. specific humidity
     b. humidity
     c. relative humidity
     d. dew point

28. When air temperature rises:
     a. condensation will likely occur
     b. the relative humidity is lowered
     c. the air can hold less water vapor
     d. the dew point is approached

29. The dew point temperature of an air parcel:
     a. is where evaporation begins to occur
     b. is where the relative humidity is one-half the
         specific humidity
     c. is approached as the air heats
     d. is approached as the air cools

30. Unstable air:
    a. has a tendency to sink
    b. has a tendency to rise
    c. will undergo adiabatic heating
    d. will rarely form clouds

31. Stable air:
    a. has a tendency to sink
    b. has a tendency to rise
    c. will undergo adiabatic cooling
    d. generally results in heavy precipitation

32. Dew and frost:
    a. occur mainly when the night is clear and calm
    b. are formed by radiational heating
    c. both fall from the atmosphere
    d. are both formed by the process of sublimation

33. Radiation fog:
    a. is not associated with the formation of dew and frost
    b. is formed mainly over water surfaces
    c. is the most common fog type over land
    d. is carried mainly by wind motion

34. Advection fog:
    a. is the most common fog type over land
    b. is formed only on calm, clear nights
    c. forms when moist air flows over a colder surface
    d. only rarely forms over coastal areas of California

35. All cirrus clouds are classified as:
    a. high clouds
    b. middle clouds
    c. low clouds
    d. clouds of vertical extent

36. Air that is rising over a mountain range is undergoing:
    a. cyclonic lifting
    b. frontal lifting
    c. orographic lifting
    d. convectional lifting

TESTING: SHORT ANSWER AND ESSAY

1. Discuss what changes occur in cloudiness and
    precipitation as an air mass flows over a mountain
    barrier. What type of lifting is involved? What
    adiabatic processes occur?   How does the "rainshadow
    effect" relate to this process?

2. Compare the causes of the precipitation patterns and the
    rough annual amounts between the equatorial zone and the
    polar zones of the earth.

3. Since polar regions have such low humidity and low
    precipitation amounts, why are there thick ice deposits
    atop such land masses as Greenland and Antarctica?

4.   What is the "lake effect" of the Great Lakes? What are
      some of the resulting impacts upon winter snowfall?

5.   Compare the environmental conditions which lead to the
      formation of radiation fog to those which lead to the
      formation of advection fog. List the main areas of North
      America where advection fogs are found most commonly.

6.   What precipitation factors other than mean annual amounts
      received have important implications to the human
      inhabitants of a region?

PLACE AND PHYSICAL FEATURE LOCATION

THE SOVIET UNION MAP EXERCISE

Locate the following by name or number on the map of the **Soviet Union** (NOTE: you may wish to sketch in the major landforms).

SOVIET REPUBLICS

1. Latvia (former SSR)
2. Lithuania (former SSR)
3. Estonia (former SSR)
4. Ukrainian SSR
5. Russian Soviet Federated Socialist Republic
6. Belorussian SSR
7. Georgian SSR
8. Armenian SSR
9. Azerbaijan SSR
10. Uzbeck SSR
11. Turkmen SSR
12. Kazakh SSR
13. Kirghiz SSR
14. Modavian SSR
15. Tadzhik SSR

* SSR = Soviet Socialist Republic (Sept., 1991)

MAJOR CITIES

16. Vladivostok
17. Kiev
18. Gorky
19. Novosibirsk
20. St. Petersburg (Leningrad)
21. Donetsk
22. Moscow
23. Odessa
24. Baku
25. Irkutsk
26. Minsk
27. Verkhoyansk
28. Volgograd
29. Omsk
30. Sverdlovsk
31. Archangelsk

PHYSICAL FEATURES

32. Kamchatka Peninsula
33. Sea of Okhotsk
34. Don River
35. Caucasus Mountains
36. Sakhalin
37. Black Sea
38. Ural Mountains
39. West Siberian Lowland
40. White Sea
41. Sea of Japan
42. Barents Sea
43. Lake Baykal
44. Volga River
45. Amur River
46. Caspian Sea
47. Ob River
48. Irtysh River
49. Lena River
50. Lake Ladoga
51. Aral Sea
52. Lake Balkhash
53. Kola Peninsula
54. Dneiper River
55. Kyzyl Kum Desert
56. Kara Kum Desert

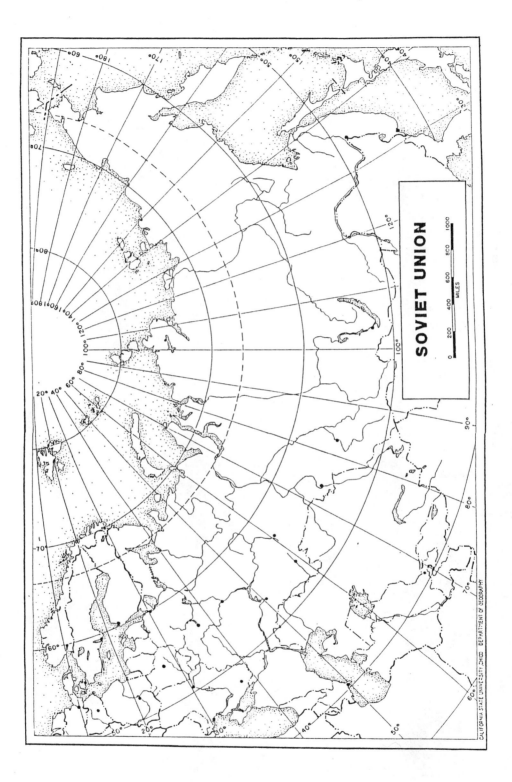

SOVIET UNION

C H A P T E R   8

WEATHER SYSTEMS

LEARNING OBJECTIVES

1. Discuss the concept of air mass formation in specific source regions.

2. Describe the temperature and moisture characteristics of the four basic types of air masses.

3. Explain the latitudinal relationship that exists in a consideration of air mass formation.

4. Differentiate between traveling anticyclones and cyclones and their wind circulations.

5. Define a weather front, explain how weather fronts are classified, and describe the four major types of fronts.

6. Describe the sequence of weather events that occur with the passage of a cold front and a warm front.

7. Outline the development and life cycle of a typical frontal cyclone and discuss the sequence of weather conditions associated with these storm systems.

8. Describe the conditions which lead to the development of thunderstorms and the weather events associated with these systems.

9.   Compare and contrast air mass thunderstorms and frontal thunderstorms in terms of origins and weather events.

10. Describe the conditions which lead to the formation of a tornado and discuss these storm systems in terms of their meteorological impacts.

11. Compare and contrast the tropical weather systems known as easterly waves, tropical depressions, and tropical storms and outline their impacts.

12. Explain the conditions which must occur for the formation of a hurricane and identify the major world areas where these occur.

13. Relate the meteorological effects of hurricanes as major storms with definite environmental impacts upon both human and natural systems.

KEY TERMS AND CONCEPTS

Air Mass

Air Mass Source Region

Air Mass types:

    Continental Polar

    Maritime Polar

    Continental Tropical

    Maritime Tropical

    Continental Arctic

    Equatorial

Anticyclone

Cyclone

Weather front

Polar Front

Frontal Cyclone

Cold Front

Warm Front

Stationary Front

Occluded Front

Air Mass Thunderstorm

Frontal Thunderstorm

Squall Line

Tornado (funnel cloud)

Easterly Wave

Tropical Depression

Tropical Storm

Hurricane

Typhoon

Storm Surge

TESTING: TRUE-FALSE

_____   1. Most air mass source regions are in the middle
              latitudes.

_____   2. A traveling anticyclone is a moving mound of air.

_____   3. A weather front is the narrow boundary zone
              between adjacent air masses.

_____   4. The polar front is a boundary separating air
              masses of arctic and subpolar origin (cA and cP
              air).

_____   5. There is a strong shift from northwest to
              southwest winds after the passage of a cold front.

_____   6. Thunderstorms are characteristically formed along
              warm fronts.

_____   7. Frontal cyclones typically form along the eastern
              flank of a trough in the jet stream.

_____ 8. Frontal cyclones move along the intertropical front of converging trade winds in both hemispheres.

_____ 9. Air mass thunderstorms are strictly convectional in origin.

_____ 10. Frontal thunderstorms are sometimes formed in a linear fashion along cold fronts -- a feature called a squall line.

_____ 11. A tornado is a mass of downward whirling air.

_____ 12. A hurricane is a considerably more destructive storm system than an easterly wave.

TESTING: MATCHING

A. CONTINENTAL POLAR   _____ 13. Occurs when a cold front catches up with the warm front

B. ARCTIC

C. MARITIME POLAR   _____ 14. Lined upon along a squall line

D. MARITIME TROPICAL
                          _____ 15. A cold, dry air mass
E. CONTINENTAL TROPICAL

                          _____ 16. A front associated with heavy thunderstorm activity
F. POLAR FRONT

G. COLD FRONT   _____ 17. Extended periods of steady precipitation

H. WARM FRONT
                          _____ 18. A named storm with wind speeds between 40 and 75 miles per hour
I. OCCLUDED FRONT

J. TORNADO   _____ 19. The smallest of the world's major storm types

K. EASTERLY WAVE
                          _____ 20. A warm, moist air mass
L. TROPICAL STORM

M. HURRICANE   _____ 21. The same as a typhoon

N. TROPICAL DEPRESSION_____ 22. Very weak tropical weather disturbances

O. THUNDERSTORM
                          _____ 23. A warm, dry air mass

                          _____ 24. An extremely cold cP air mass

TESTING: MULTIPLE CHOICE

25. Source regions for air mass formation are:
    a. located in the high and low latitudes
    b. located in the middle latitudes
    c. located only over the continents
    d. always areas with air having high moisture content

26. Continental Polar air masses:
    a. form over the North Pacific Ocean
    b. are warm and dry
    c. are cool and moist
    d. form over Northern Canada during winter

27. Maritime Polar air masses:
    a. form over the polar ice sheets
    b. form over subpolar ocean surfaces
    c. are not associated with frontal cyclones
    d. are cool and dry

28. Maritime Tropical air masses:
    a. are warm and dry
    b. are cool and moist for their latitude
    c. are formed over tropical desert surfaces
    d. are formed over tropical/subtropical ocean surfaces

29. Continental Tropical air masses:
    a. form over continental interiors in the dry subtropics
    b. form over Siberia during the winter
    c. are warm and moist
    d. are associated with the trade winds of both
        hemispheres

30. Which front has warm temperatures before passage with
    colder temperatures after passage:
    a. warm front
    b. cold front
    c. occluded front
    d. polar front

31. An occluded front:
    a. is an areas of dense overcast
    b. is an area of clear weather conditions
    c. is not associated with frontal cyclones
    d. is the same as a polar front

32.   Along a warm front:
      a.   air mass occlusion never occurs
      b.   cool air is displacing cold air
      c.   cold air is displacing warm air
      d.   warm air is displacing cool air

33.   A tornado:
      a.   always begins as a funnel cloud over water
      b.   is associated with the base of a dense cumulonimbus
           cloud
      c.   always follows a distinctive path
      d.   is larger in size than a hurricane

34.   Easterly waves:
      a.   cannot develop into hurricanes
      b.   occur mainly along the polar fronts
      c.   are weak tropical weather disturbances in the form of
           a trough
      d.   are more intensive storm systems than hurricanes

35.   Hurricanes:
      a.   always form over tropical/subtropical ocean surfaces
      b.   usually form along the polar front
      c.   are also called tropical depressions
      d.   can develop into easterly waves

36.   Which of the following is not included on a weather map:
      a.   air pressures are given in millibars
      b.   air pressures are shown by isobars
      c.   areas of current precipitation are shaded
      d.   weather fronts are shown as dotted lines

TESTING: SHORT ANSWER AND ESSAY

1.   How are air masses formed? What happens to them as they
     are transported to other regions?

2.   What weather changes occur as a typical frontal cyclone
     passes through a region? For your answer utilize the
     five categories of weather changes listed in the text.

3.   What relationships exist between the weather systems
     known as easterly waves, tropical depressions, tropical
     storms, and hurricanes? How can each of these be
     discriminated from the others?

4.   What conditions lead to the development of tornadoes?

5.   Compare the potential for damage between a tornado and a hurricane. In what ways do these two intense storm systems differ in their possible damaging impacts?

6.   In what ways does Hurricane Hugo (1989) demonstrate the potential for disaster that can occur with a massive hurricane that moves onto the southeastern United States?

7. How can the patterns of air masses and fronts depicted on
    a weather map be used in weather forecasting?

8. Compare the following storm systems with each other:
    hurricanes, chubascos, typhoons, and "cyclones" of the
    Indian Ocean.

9. Compare the air mass interaction along the Polar Front
    with the air mass interaction in the ITCZ.

PLACE AND PHYSICAL FEATURE LOCATION

THE MIDDLE EAST MAP EXERCISE

Locate the following by name or number on the map of the
Middle East. (NOTE: you may wish to sketch in the major
landforms).

COUNTRIES

1. United Arab Emirates
2. Pakistan
3. Yemen
4. Syria
5. Iran
6. Jordan
7. Oman
8. Iraq
9. Israel
10. Cyprus
11. Egypt
12. Kuwait
13. Afghanistan
14. Lebanon
15. Yemen (PDR)
16. Turkey
17. Saudi Arabia
18. Qatar
19. Bahrain
20. Libya
21. Sudan

PHYSICAL FEATURES

37. Mediterranean Sea
38. Jordan River
39. Khyber Pass
40. Tigris River
41. Gulf of Oman
42. Lake Nasser
43. Arabian Desert
44. Red Sea
45. Euphrates River
46. Dead Sea
47. Nile River
48. Strait of Hormuz
49. Gulf of Suez
50. Persian Gulf
51. Black Sea
52. Plateau of Iran
53. Hindu Kush
54. Gulf of Aden
55. Caspian Sea
56. Suez Canal
57. Sinai Peninsula
58. Socotra
59. Rub al Khali

MAJOR CITIES

22. Mecca
23. Damascus
24. Beirut
25. Baghdad
26. Riyadh
27. Jerusalem
28. Alexandria
29. Amman
30. Tehran
31. Kabul
32. Tripoli (Libya)
33. Tel Aviv
34. Cairo
35. Khartoum
36. Ankara

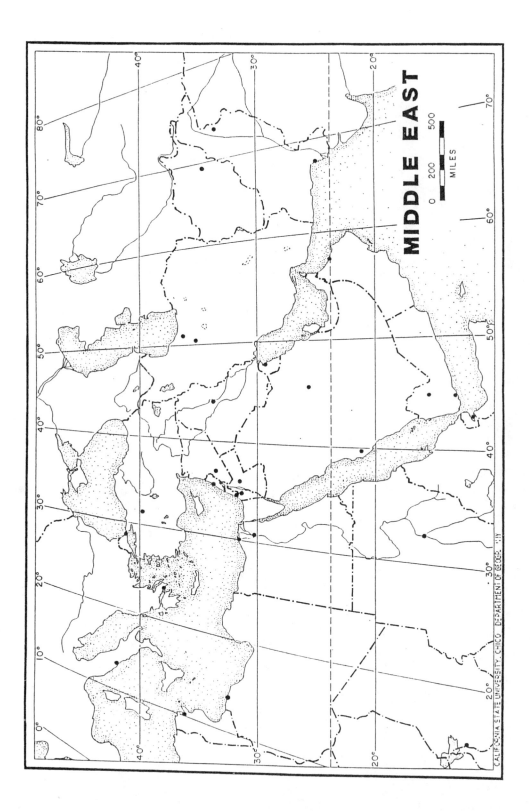

MIDDLE EAST

0   200   500
MILES

C H A P T E R   9

CLIMATES OF THE WORLD

1. Describe the chief advantages and limitations of climate classification systems.

2. Discuss the main climatic factors and the major reasons for the existence of the two main humid tropical climates - the Tropical Wet Climate and the Tropical Wet and Dry Climate.

3. Outline the main climatic features and the major reasons for the existence of the Low Latitude Dry Climates.

4. Describe the main climatic features and the major reasons for the existence of the four humid middle latitude climates - the Dry Summer Subtropical Climate, the Humid Subtropical Climate, the Marine Climate, and the Humid Continental Climates.

5. Discuss the main climatic features and the major reasons for the existence of the Mid Latitude Dry Climate.

6. Discuss the climatic features and the major reasons for the existence of the three High Latitude Climates - the Subarctic Climate, the Tundra Climate, and the Polar Climate.

7. Discriminate the Highland Climates from the other major world climates on the basis of major climatic features and the impact of the complexity of upland areas.

8.   Categorize the types of climatic changes that have occurred over urban areas due to human activities.

9.   Diagram the basic method of depicting climatic data using the climagraph.

10.   Outline the evidence for and the basic patterning of climatic fluctuations during the past 10,000 years of earth history.

11.   Discuss the specific advantages and disadvantages of the Koppen Climatic Classification System.

KEY TERMS AND CONCEPTS

Low Latitude Climates

Humid

Arid

Semiarid

Middle Latitude Climates

High Latitude Climates

Highland Climates

Urban Heat Island

Climagraph

TESTING: TRUE/FALSE

_____   1.   Regional climatic analysis can easily be made without the compilation of past weather data.

_____   2.   The differences between the three major climatic types of the low latitudes are created by the placement of the subtropical highs and the Intertropical Convergence Zone.

_____   3.   The seasonal differences in temperature are greater than the daily ranges in the Tropical Wet Climate.

_____   4.   Summer is the dry season in regions of Tropical Wet and Dry Climate.

____   5.   In the dry climates, potential evapotranspiration
             is always less than precipitation amounts.

____   6.   The Humid Subtropical Climate is also called the
             Mediterranean Climate due to its extensive develop-
             ment around the coasts of that water body.

____   7.   The poleward shift of the subtropical high is
             responsible for the dry season of the Dry Summer
             Subtropical Climate.

____   8.   The Humid Subtropical Climate is always found on
             the middle latitude west coasts of the continents.

____   9.   The Humid Subtropical Climatic areas receive sub-
             stantial amounts of precipitation throughout the
             year.

____   10.   Most Mid Latitude Dry climatic areas are found in
             rainshadow locations and in mid-continental regions
             isolated from oceanic moisture sources.

____   11.   Mid Latitude Dry climates differ mainly from Low
             Latitude Dry climates on the basis of temperature
             values rather than precipitation totals.

____   12.   Marine climatic areas are primarily found on the
             eastern coasts of the continents between 40 and 60
             degrees N and S latitude.

____   13.   There are no areas of Marine climate in the
             Southern Hemisphere due to the absence of large
             continental land masses in the high middle
             latitudes.

____   14.   Although precipitation totals are modest, there is
             no dry season in areas of Humid Continental
             Climate.

____   15.   The Subarctic climate exists only in the northern
             interiors of North America and Eurasia.

____   16.   The world's largest annual temperature ranges are
             found in areas of Polar climates.

____   17.   In areas of Tundra climate, the mean temperature of
             the warmest month is between 32 and 50 degrees F.

____   18.   Climatic conditions in the world's highland areas
             are quite homogeneous which allows them to be
             conveniently grouped together as Highland climates.

_____ 19. A significant impact of cities on their local
            environments is to create islands of urban cooling.

_____ 20. One advantage of the climatic classification system
            used in the text is its simplicity inasmuch as it
            is primarily based on basic annual temperature and
            precipitation data.

TESTING: MATCHING I

(Each climatic name can be used more than once)

Climatic Names: **Koppen Symbols:**

A. TROPICAL WET   _____ 21. Dfa, Dfb, Dwa and Dwb

B. TROPICAL WET AND DRY   _____ 22. ET

C. LOW LATITUDE DRY   _____ 23. Bwk and Bsk

D. DRY SUMMER SUBTROPICAL   _____ 24. Cfb and Cfc

E. HUMID SUBTROPICAL   _____ 25. Af and Am

F. MID LATITUDE DRY   _____ 26. Dfc, Dfb, Dwc and Dwd

G. MARINE   _____ 27. EF

H. HUMID CONTINENTAL   _____ 28. H

I. SUBARCTIC   _____ 29. Bwh and Bsh

J. TUNDRA   _____ 30. Csa and Csb

K. HIGHLAND   _____ 31. Aw

L. POLAR

TESTING: MATCHING II

Climatic Name:

A. TROPICAL WET

B. TROPICAL WET AND DRY

C. LOW LATITUDE DRY

D. DRY SUMMER SUBTROPICAL

E. HUMID SUBTROPICAL

F. MID LATITUDE DRY

G. MARINE

H. HUMID CONTINENTAL

I. SUBARCTIC

J. TUNDRA

K. HIGHLAND

L. POLAR

Major World Regions:

____ 32. Central Chile, Southern Europe, and Capetown, South Africa

____ 33. Southern Japan, Eastern China, and Uruguay

____ 34. Southwestern United States, the Atacama of Peru and Chile

____ 35. Central Asia, Patagonia, and the Great Plains of North America

____ 36. South coasts of Alaska, Chile, and New Zealand

____ 37. Amazon and Congo Basins

____ 38. Coastal Greenland, Northern Alaska, the north coast of Eurasia

____ 39. Vietnam, Northeastern Australia, Indonesia, the Philippines

____ 40. The Sahara and the Kalahari

____ 41. Southern Florida and Southern Brazil

____ 42. Central California and Southern and Southeastern Australia

____ 43. Western Norway, Western Europe, and the British Isles

____ 44. Interior Greenland and Antarctica

_____ 45. Central Alaska,
           Scandinavia, Northern
           interior of the Soviet
           Union

_____ 46. New England, Eastern
           Europe, and North Korea

TESTING: MULTIPLE CHOICE

47. The most equatorward of the earth's climatic types if
    the:
    a. Tropical Wet and Dry Climate
    b. Humid Subtropical Climate
    c. Tropical Wet Climate
    d. Low Latitude Dry Climate

48. In the text's climatic classification system "B" climates
    are:
    a. Polar climates
    b. Dry climates
    c. Middle Latitude climates
    d. Humid Continental climates

49. The Tropical Wet and Dry Climate has:
    a. the ITCZ in summer and the subtropical high in winter
    b. the Polar Front in winter and the subtropical high in
        summer
    c. the ITCZ all year round
    d. the Polar Front all year round

50. The Low Latitude Dry Climate has:
    a. east coast locations
    b. frontal cyclone activity in summer only
    c. influence of the subtropical high in winter only
    d. cold ocean currents offshore

51. The greatest daily ranges of temperature are found in
    the:
    a. Tundra Climate
    b. Tropical Wet Climate
    c. Subarctic Climate
    d. Low Latitude Dry Climate

52. The Dry Summer Subtropical Climate is:
    a. classified as Csa-Csb in the Koppen system
    b. found only in North America and Eurasia
    c. a climate with no summer temperature above $50^\circ F$
    d. dominated by the ITCZ in the summer only

53. Humid Subtropical climatic areas have:
    a. mT air in summer and weather fronts in winter
    b. weather fronts all year round
    c. a strong summer dry season
    d. ITCZ activity in summer and cP air in winter

54. The Mid Latitude Dry Climates are:
    a. only located in the Northern Hemisphere
    b. centered on the Tropics of Cancer and Capricorn
    c. roughly 90 percent of the earth's dry climatic areas
    d. cold winter dry climates

55. Frontal cyclonic storms are both numerous and well-developed in the:
    a. Tropical Wet Climate
    b. Subtropical Humid Climate
    c. Subarctic Climate
    d. Marine Climate

56. The Marine Climate is the dominant climatic type of:
    a. North America
    b. Europe
    c. The Mediterranean Sea Basin
    d. Australia

57. The mean annual percentage of cloudiness and the total duration of cloudiness is highest in the:
    a. Marine Climate
    b. Subarctic Climate
    c. Tundra Climate
    d. Tropical Wet and Dry Climate

58. The Humid Continental Climates have:
    a. a North American and Eurasian distribution
    b. a strong summer dry season
    c. summer frontal cyclonic storm activity
    d. a very small annual temperature range

59. The Subarctic Climate is:
    a. found only in Antarctica and Greenland
    b. a very severe winter climatic type
    c. a very rainy, cloudy climatic type
    d. dominated by very low winter pressures

60. Areas of Tundra Climate:
    a. have no average monthly temperature above freezing
    b. are not found in the Southern Hemisphere
    c. are located along subpolar coasts
    d. are too cold to have any vegetation

61. The Polar Climate:
    a. is found only in Antarctica
    b. has only four months with average temperatures above 50°F
    c. is one of the rainiest climatic types
    d. receives minimal frontal cyclonic precipitation which is virtually all frozen

62. Highland Climates:
    a. are very similar from place to place over the earth
    b. are all classified as dry climates due to elevation
    c. are a complex mosaic of diverse climatic types
    d. are virtually indistinguishable from the climates of adjacent lowland areas

63. Urban climates differ from those of rural areas in the following variable:
    a. temperature
    b. wind patterns
    c. atmospheric moisture
    d. all of the above

TESTING: SHORT ANSWER AND ESSAY

1.   What are the main advantages and disadvantages of
     climatic classification systems such as the system used
     in the text?

2.   In what major ways is the Low Latitude Dry Climate
     similar to the Mid Latitude Dry Climate? In what major
     ways, in both origin and location, are they different?

3.   Why would the Tropical Wet and Dry Climate and the
     Subtropical Summer Dry Climate be difficult climates for
     agriculture?

4.   Why would the Marine Climate, the Humid Subtropical
     Climate, and the Humid Continental Climates be
     especially suitable for agriculture? Do the three
     humid climates have any drawbacks for crop growing?

5.   What are the main reasons for having three distinctive
     climatic types within the tropics that are basically
     separated on the basis of season and amount of
     precipitation? What are the three climatic types?

6. What are the main climate variables that interact to give
    <u>Highlands</u> different climatic conditions than those of
    adjacent lowlands?

7. How have human activities in urban areas created
    definite, observable climatic changes in these areas?
    What processes are involved?

8. How and why is a "climagraph" used in climatic
    depictions? What types of climatic data are displayed on
    a climagraph?

PLACE AND PHYSICAL FEATURE LOCATION

EAST ASIA MAP EXERCISE

Locate the following by name or number on the map of **East Asia**. (NOTE: you may wish to sketch in the major landforms).

POLITICAL DIVISIONS

1. Japan
2. South Korea
3. Hong Kong
4. North Korea
5. Mongolia
6. Macao
7. People's Republic of China

PHYSICAL FEATURES

8. Taiwan (Formosa)
9. Philippine Sea
10. Gobi Desert
11. South China Sea
12. Huang or Yellow River
13. Amur River
14. Korea Strait
15. East China Sea
16. Sakhalin
17. Yellow Sea
18. Xun Xi River
19. Szechwan Basin
20. Yangtze or Chang River
21. Taiwan Strait
22. Plateau of Tibet
23. Sea of Japan
24. Hokkaido
25. Honshu
26. Kyushu
27. Shikoku
28. Ryukyu Islands
29. Manchuria
30. Sichuan
31. Takla Makan Desert
32. Tarim Basin
33. Kurile Islands

MAJOR CITIES

34. Seoul
35. Osaka
36. Shanghai
37. Yokohama
38. Kyoto
39. Beijing (Peking)
40. P'yongyang
41. Pusan
42. Harbin
43. Tokyo
44. T'aipei
45. Kobe
46. Chengdu
47. Canton
48. Hiroshima
49. Ulan Bator
50. Sapporo
51. Victoria (Hong Kong)

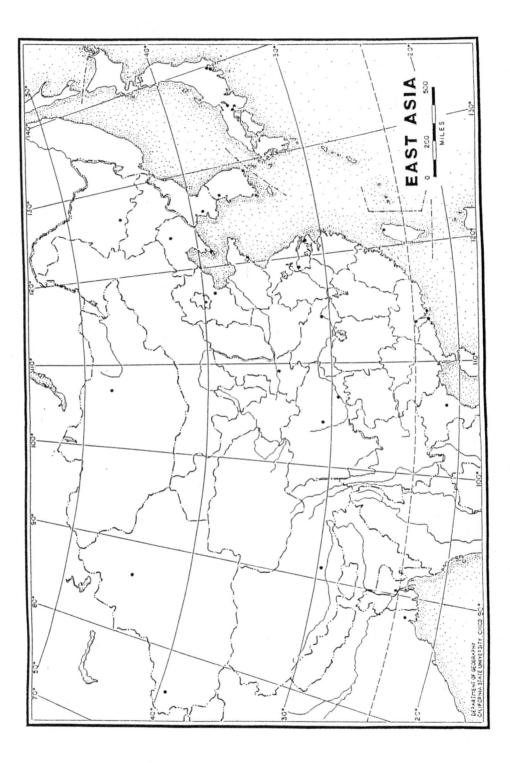

EAST ASIA

C H A P T E R   10

WATER ON AND BENEATH THE EARTH'S SURFACE

LEARNING OBJECTIVES

1. Acquaint yourself with the movement of water through the subsurface of the earth; this is a part of the landscape of physical geography which cannot be directly perceived.

2. Examine the basic motion of water through the soil materials and into the subsurface. Relate this movement to plant growth and stream level.

3. Inspect the terminology associated with groundwater and develop a sensitivity to the problems associated with this important water resource.

4. Insure that you are familiar with the distribution of fresh water, salt water, groundwater, and ice on the earth's surface. Develop a sense of what proportion of water is contained in each source.

5. Relate perennial streams and intermittent stream to changes in the water table.

6. Examine closely the origin and evolution lakes in order to develop a sensitivity for the uniqueness of the occurrence of this aquatic landform type.

7. Survey the distribution of lakes around the world and establish the significant contribution they make to societies relying upon them for water, transportation, recreation, and industrial uses.

8.   Explore the distribution of rivers and streams around
     the world and examine the variability of discharge. All
     discussions of surface water can be strongly correlated
     to the climate of an area.

9.   Examine the concept of the stream regime.

10. Become aware that many of the great population centers of
     the world are located along the floodplains of major
     rivers. An awareness of the relationship of the flood-
     plain to the stream regime is a valuable tool for looking
     at human landscapes.

11. Investigate the physical and chemical nature of the
     oceans.

12. Examine the origin and nature of the different types of
     oceanic ice: pack ice, sea ice, and ice bergs.

13. Analyze the horizontal circulation of the oceans and
     develop an awareness of how this alters climates.

14. Look at the processes resulting in ocean wave and tidal
     movements.

KEY TERMS AND CONCEPTS

Provide a short definition or description of the following key
words and concepts from the chapter.

Underground Water

Soil Water

Zone of Aeration

Zone of Saturation

Ground Water

Water Movement

Interception

Porosity

Permeability

Throughflow

Water Table

Deep Water

Aquifer

Aquicludes

Fossil Water

Streams

Stream Discharge

Spring

Perennial Streams

Ephemeral Streams

Stream Regime

Exotic Streams

Floods

Crests

Floodplains

Drainage Basin

Lakes

Origins

Crustal Movements

Glacial Erosion and Deposition

Solution Depressions

Volcanic Actions

Mass Movements

Human Impoundments

Swamps

Water Distribution

Fresh Water

TESTING: TRUE/FALSE

_____   1. The quantity of fresh water in lakes and rivers is
             less than 1/77 of that present in groundwater.

_____   2. Groundwater which is less than one-half mile from
             the surface of the earth is generally available
             for extraction.

_____   3. The earth's dominant river in all respects except
             volume is the Nile River.

_____   4. More water is held in glacial ice than in ground-
             water and fresh water combined.

_____   5. The two basic criteria for a lake are a surface
             depression and the availability of water to fill
             it.

_____   6. The zone of aeration lies immediately below the
             zone of saturation.

_____   7. Ocean water generally contains about 0.35% salts
             by weight.

_____   8. The driving factor behind surface ocean currents
             is the wind; the driving factor behind deep
             currents is density differences.

_____   9. The Humboldt current is generally located off of
             the coast of Baja California and affects our
             weather during the El Nino years.

_____   10. "Fossil water" is a groundwater supply that is
             replenished by each year's precipitation.

_____   11. The water levels of the five Great Lakes are
             rising.

_____   12. The Ogallala Aquifer of the U.S. High Plains is
             being rapidly replenished by precipitation.

TESTING: MATCHING

13. ____ fetch   A. broken thin ice cover

14. ____ glacial ice   B. combination of tide and river

15. ____ neap tides   C. origin in wind currents

16. ____ landslides   D. ranks after groundwater

17. ____ tidal currents   E. related to lunar gravity

18. ____ fresh water   F. coastal rising of water

19. ____ pack ice   G. distance wind blows

20. ____ surface currents   H. possible source of a lake

21. ____ upwelling   I. small tidal range

22. ____ tidal bore   J. scarce global resource

TESTING: MULTIPLE CHOICE

Choose the best response for each of the following multiple
choice questions. Each question has only one correct answer.

23. The proportion of water held in the atmosphere is:
    a. 0.0001%
    b. 0.10%
    c. 1.0%
    d. 10%

24. A type of tectonic activity which can result in the
    development of a lake produces a landform called a:
    a. graben
    b. fault
    c. synclinal valley
    d. cirque

25. Of all the precipitation falling on the land, the
    proportion moving to the oceans is thought to be:
    a. 10%
    b. 30%
    c. 45%
    d. 60%

26. The zone of permanently saturated soil or rock materials
    in which all pores are filled with water is called:
    a. the zone of aeration
    b. the swamp zone
    c. the water table
    d. the local aquifer

27. The factor which does not help to determine wave size is:
    a. the fetch
    b. the direction the wind is blowing
    c. the velocity of the wind
    d. the length of time the wind has been blowing

28. In normal years the water in the Humboldt Current off of
    the coast of Peru is:
    a. cold, upwelling water
    b. cold, sinking water
    c. warm, upwelling water
    d. warm, sinking water

29. The largest freshwater lake in the world is:
    a. the Caspian Sea
    b. Lake Superior
    c. Lake Victoria
    d. Lake Huron

30. The regime of a stream is its
    a. flood hydrograph
    b. the seasonal variations in discharge
    c. the character of the area through which it flows
    d. listing of plant and animal life

31. Icebergs:
    a. make up the Arctic ice cap
    b. result from ice broken off glaciers entering the
        ocean
    c. are seldom more than a few tens of feet thick
    d. present no hazard to modern shipping

32. Spring tides:
    a. have a low range
    b. result from the influence of the moon only
    c. reflect the interaction of the gravitational pull
        of the sun and the moon in alignment
    d. occur near the vernal equinox

33. About 96.7 percent of all of the water of the world is
     located in:
     a. underground reservoirs
     b. surface lakes and streams
     c. glacial ice
     d. none of the above answers is correct

34. Fossil water:
     a. contains minerals leached from the bones of dinosaurs
     b. has its origin in past wet climates
     c. is not a resource for agriculture
     d. is being replenished by contemporary rainfall

35. The surface runoff from an intense storm is:
     a. relatively rapid
     b. takes a long time to reach the streams
     c. rises slowly and falls rapidly
     d. inundates the floodplain of the stream

36. Although sea ice movements are primarily controlled by
     the wind, the movement of icebergs is controlled by:
     a. topographic influence of surrounding mountains
     b. the wind
     c. oceanic currents
     d. coastal upwelling and density currents

37. An ion is:
     a. an electrically charged atom or group of atoms
     b. an electrically neutral atom or group of atoms
     c. a type of salt crystal present in the ocean
     d. the source of saltiness in fresh water

TESTING: SHORT ANSWER AND ESSAY

1.   What main factors will interact to affect the groundwater
     supplies of an area?

2.   How do lakes form and disappear?   Why are lakes so
     infrequently encountered on the surface?

3.   Make a list of the principal chemical and physical
     characteristics of ocean water.   What is the origin of
     these characteristics?

4. Describe the movement of a water droplet falling as rain, and finally flowing into a perennial stream.

5. Name the causes of the motion of ocean tides. Name the causes of the motion of ocean currents. Name the causes of ocean waves.

6. Describe how stream discharge can be measured.

7.   What are the regional and global implications of the
     phenomenon known as El Nino?

8.   In what ways can human activities impact upon groundwater
     supplies and their quality? How does the human use of
     the Ogallala Aquifer relate to this question?

PLACE AND PHYSICAL FEATURE LOCATION

SOUTH ASIA MAP EXERCISE

Locate the following by name or number on the map of **South Asia**. (NOTE: you may wish to sketch in the major landforms).

COUNTRIES

1. India
2. Pakistan
3. Afghanistan
4. Sri Lanka
5. Nepal
6. Bhutan
7. Bangladesh
8. Burma

MAJOR CITIES

18. Kabul
19. Lahore
20. Karachi
21. New Delhi
22. Bombay
23. Calcutta
24. Dacca
25. Kathmandu
26. Colombo
27. Rangoon
28. Islamabad

PHYSICAL FEATURES

9. Himalaya Mountains
10. Bay of Bengal
11. Deccan Plateau
12. Indus River
13. Ganges River
14. Brahmaputra River
15. Arabian Sea
16. Andaman Sea
17. Malabar Coast

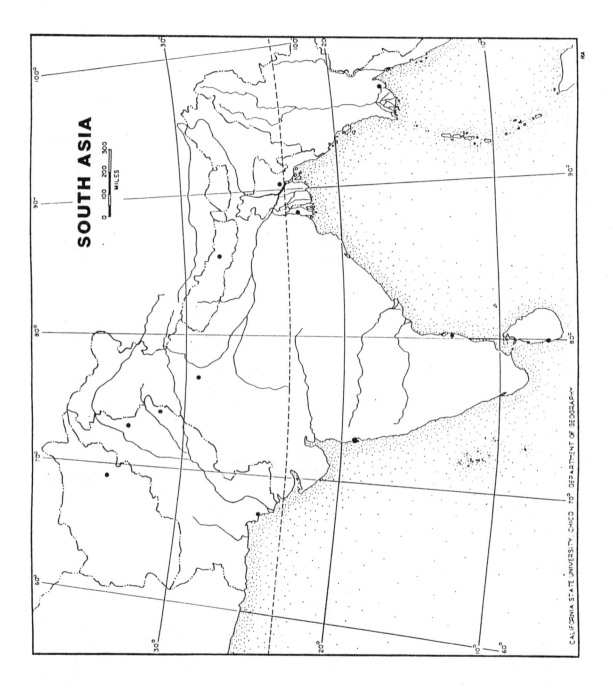

SOUTH ASIA

C H A P T E R   11

BIOGEOGRAPHY

LEARNING OBJECTIVES

1. List and explain the four major biological controls on the distribution of life forms.

2. Describe the primary ecological processes which affect the distribution of localized biological communities.

3. Outline the basic framework of vegetative succession.

4. Understand the basic ecological aspects of animals and plants occupying the same ecosystem.

5. Discuss the four major associations of natural vegetation and relate these to their particular range of environmental conditions.

6. Outline the major environmental factors and the ways these interact to influence the growth of vegetation.

7. Identify the range of moisture impacts upon natural vegetation that is illustrated by xerophytes and hygrophytes at both extremes.

8. List the eight major forest associations and differentiate between them in terms of structure and major world regions that they occupy.

9. Compare and contrast the environmental interrelationships of broadleaf and needleleaf trees.

10. Compare and contrast the climatic and community interrelationships of evergreen and deciduous forest types.

11. Discuss the reasons for dividing the world's forests into tropical and nontropical associations and identify the zones of contact between them.

12. Explain the major natural vegetation associations of North America and be able to discuss the environmental factors which create this patterning.

13. Identify the forest associations which would be encountered as one followed a transect from the American Midwest, across the Great Lakes, through Northern Canada to the coast of the Arctic Ocean.

14. Describe the three major grassland associations and differentiate between them in terms of structure and major world regions that they occupy.

15. Discuss the controversy regarding human activities in altering world grassland associations.

16. Describe the basic attributes of desert vegetation and identify the most widespread botanical groupings of desert plants.

17. Discuss the major environmental factors which control the tundra vegetation and relate this association with the neighboring taiga.

18. Outline the process whereby the earth's forests, and especially the tropical rainforests, are being destroyed and discuss the environmental consequences of this deforestation.

19. Identify the environmental impacts of acid rain and especially as this phenomenon affects the United States, Canada and Europe.

20. Provide a balanced discussion of the mesquite invasion in the American southwest during the past century in terms of causative factors and distinctive consequences.

21. Become conversant with the major biological and other environmental consequences of acid rain in those areas most affected.

KEY TERMS AND CONCEPTS

Biogeographic Characteristics

Photosynthesis

Biogeography

Biological Community

Ecosystem

Ecotone

Mutation

Macrohabitat

Microhabitat

Herbivore

Carnivore

Vegetation

Primary Succession

Climax Community

Plant Associations

Biomass

Barriers

Moisture Influences

Xerophytes

Hygrophytes

Mesophytes

Tropophytes

Acid Rain

Growth Characteristics

Growing Season

Evergreen

Deciduous

Perennial

Annual

Forest

Broadleaf

Needleleaf

Conifer

Epiphytes

Vegetation Types

Sclerophyll

Chaparral

Taiga

Muskeg (Tundra)

Savanna

Steppe

Woodland

Prairie

Halophytes

Ephemerals

Riparian

TESTING: TRUE/FALSE

_____ 1. A plant and animal assemblage whose members occupy a particular environmental setting is known as an ecosystem.

_____ 2. A transitional zone between ecological communities is known as an ecotone.

_____ 3. Primary succession of vegetation is regrowth after an episode of vegetation clearing such as a forest fire or over-grazing.

_____ 4. An herbivore is a meat-eating animal.

_____ 5. All mutations are beneficial for biotic populations.

_____ 6. A specific site with appropriate environmental conditions for a species is termed a habitat.

_____ 7. Unfavorable areas between habitats can become "barriers" to dispersal.

_____ 8. The availability of moisture is the single most important factor controlling large-scale patterns of natural vegetation.

_____ 9. Deciduous trees differ from evergreen trees in their ability to shed their leaves during unfavorable seasons.

_____ 10. Both annuals and perennials have multiple-year lifespans.

_____ 11. Vegetative patterns tend to be more diverse in rugged areas than in areas of lower, flatter topography.

_____ 12. Soil fertility for plant growth is solely a function of the moisture content of the soil.

_____ 13. Oaks, maple and hickory are known as coniferous trees while firs, pines, and spruce are examples of broadleaf trees.

_____ 14. The tropical rainforest has neither a cold season nor a lengthy season of drought.

_____ 15. The floors of tropical rainforest are usually covered with a jungly layer of dense undergrowth.

_____ 16. The West Coast Coniferous Forest is an association of short, scrubby trees with very few species represented.

_____ 17. The tropical savannas are generally situated on the drier poleward margins of the Tropical Scrub Woodland.

_____ 18. Savannas, steppes and prairies are found in the same basic climatic zones.

_____ 19. Trees are not found in most desert areas because their moisture requirements are generally too high.

_____ 20. The tundra is the most cold-tolerant vegetation association.

TESTING: MATCHING

A. XEROPHYTES   _____ 21. Plants adapted to alternating favorable and unfavorable climatic conditions.

B. HYGROPHYTES

C. MESOPHYTES   _____ 22. Trees and shrubs occurring in Mediterranean Scrub associations

D. TROPOPHYTES

E. EPIPHYTES   _____ 23. Tropical grasslands with scattered trees and shrubs

F. HALOPHYTES
                        _____ 24. "Air plants" growing on tree branches
G. EPHEMERALS

H. MUSKEG   _____ 25. A stream-side association of plants

I. TAIGA
                        _____ 26. Plants adapted to dry regions
J. SAVANNA
                        _____ 27. The Northern Coniferous Forest
K. PRAIRIE
                        _____ 28. Plants occupying well-drained habitats
L. SCLEROPHYLLS

M. RIPARIAN   _____ 29. The short grasslands of the low and middle latitudes

N. STEPPE
                        _____ 30. Plants adapted to excessively moist environments

_____ 31. Salt-tolerant plants of desert regions

_____ 32. Small, short-lived desert plants

TESTING: MULTIPLE CHOICE

33. A climax community is:
    a. a functioning entity consisting of all organisms in a community
    b. a plant community adapted to changing moisture conditions
    c. the plant community best suited to survive under existing environmental conditions
    d. a transitional zone between the plant communities

34. Plants adapted for survival in arid regions are:
    a. Hygrophytes
    b. Xerophytes
    c. Epiphytes
    d. Mesophytes

35. Which of the following is not a major environmental necessity for vegetation:
    a. cloud cover
    b. moisture
    c. temperature
    d. light

36. A forest association:
    a. is found only in the tropics
    b. is an assemblage only of trees
    c. is composed only of broadleaf, deciduous trees
    d. is composed primarily of trees with overlapping foliage

37. Which of the following is not a description of the tropical rainforest:
    a. it contains eighty percent of the world's total biomass
    b. it has buttressed root systems and epiphytic plants
    c. its trees are deciduous
    d. its trees are broadleaf hardwoods

38. The Tropical Semideciduous Forest:
    a. is located on the margins of the tropical rainforest
    b. is composed of broadleaf, evergreen trees
    c. is also called the jungle
    d. is the tallest of the world's forest associations

39. The "connecting link" between the tropical and non-
     tropical forests is:
     a. the Northern Coniferous Forest
     b. the Sclerophyll Scrub Forest
     c. the Tropical Rainforest
     d. the Subtropical Broadleaf Evergreen Forest

40. The Mediterranean Woodland and Scrub:
     a. is found in areas with no dry season
     b. is also called the Chaparral
     c. is found on the margins of the tropical forests
     d. is found only around the Mediterranean Sea

41. The Mid Latitude Deciduous and Mixed Forest:
     a. is composed of coniferous trees
     b. contains broadleaf trees such as oaks, maple, and
         hickory
     c. is only found in the United States
     d. grows in areas with no cold winter season

42. Tall coastal redwoods, Douglas fir, and other conifers
     are in the:
     a. Mid Latitude Deciduous and Mixed Forest
     b. Northern Coniferous Forest
     c. West Coast Coniferous Forest
     d. Subtropical Broadleaf Evergreen Forest

43. The Northern Confierous Forest:
     a. is found in areas of Polar Climate
     b. is found in regions with a Subarctic Climate
     c. is found between prairies and sclerophyllous scrub
         associations
     d. is found only in Eurasia

44. This tall grass association occupies the smallest areal
     coverage of any world grassland:
     a. the Steppe
     b. the Prairie
     c. the Savanna
     d. the Tundra

TESTING: SHORT ANSWER AND ESSAY

1.   What are the four main global environmental controls on the distribution of natural vegetation?

2.   What are the four major associations of natural vegetation? What characteristics differentiate them?

3.   In what ways is moisture availability critical to the distribution of natural vegetation? How is this illustrated by the different growth strategies or xerophytes as contrasted with hygrophytes?

4.   What are the eight major forest associations and how can
     they be related to specific climatic types?

5.   What are the three major grassland associations and how
     can they be differentiated in terms of structure and the
     main world regions that they occupy?

6.   What are the predominant impacts of such human activities
     as deforestation and combustion leading to acid rain?
     Which regions seem to be most affected by these
     processes?

7. How have human activities caused a retreat of the earth's
    tropical rainforests? What world areas are most
    affected, and what are the chief consequences of the
    deforestation of this vegetation community?

8. What is meant by the "mesquite invasion" in the American
    Southwest? What environmental changes have led to this
    invasion, and what are the major human land use
    consequences?

PLACE AND PHYSICAL FEATURE LOCATION

SOUTHEAST ASIA MAP EXERCISE

Locate the following by name or number on the map of **Southeast Asia**. (NOTE: you may wish to sketch in the major landforms).

COUNTRIES

1. Burma
2. Brunei
3. Vietnam
4. Indonesia
5. Thailand
6. Laos
7. Philippines
8. Papua New Guinea
9. Kampuchea
10. Malaysia

MAJOR CITIES

11. Surabaya
12. Kuala Lumpur
13. Manila
14. Phnom Penh
15. Bangkok
16. Singapore
17. Ho Chi Minh City
18. Rangoon
19. Hanoi
20. Jakarta

PHYSICAL FEATURES

21. Hainan
22. Malay Peninsula
23. Celebes
24. Irrawaddy River
25. South China Sea
26. Andaman Sea
27. Mekong River
28. Indian Ocean
29. Strait of Malacca
30. Celebes Sea
31. Gulf of Tonkin
32. Gulf of Thailand
33. Philippine Sea
34. Java
35. New Guinea
36. Luzon
37. Borneo
38. Mindanao
39. Sumatra
40. Timor Sea
41. Bali

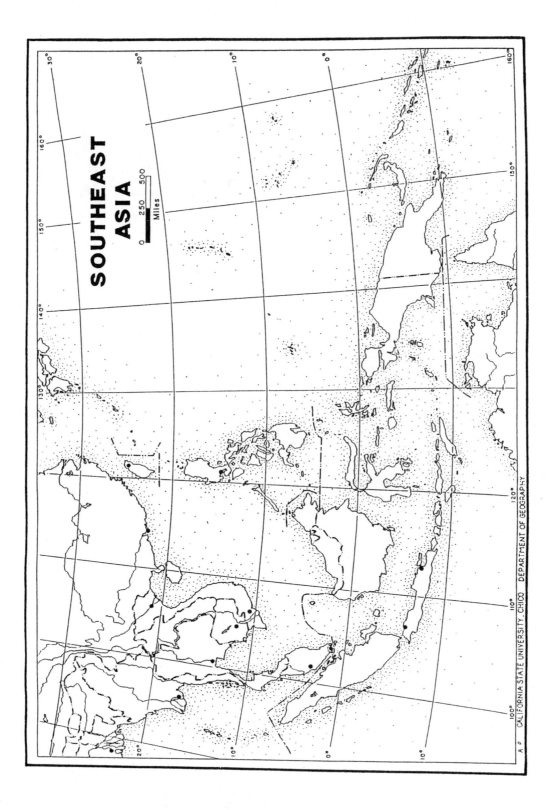

C H A P T E R   12

SOILS

1. Develop an introduction to the terminology of soils geography. The terminology is complex and unfamiliar, but must be understood if the material is to comprehended.

2. Grasp a sense of how a classification system is devised and what concepts are incorporated into it. The 7th Approximation system required considerable thought and ingenuity and understanding it will allow a basic familiarization with the ten major soil orders.

3. Relate soil formation to a series of major factors; these include climate, organisms, relief, parent material, and time.

4. Familiarize yourself with the mineral, organic, atmospheric, and aqueous components of the soil. The soil is a complex system which is difficult to fully appreciate.

5. Relate common soil characteristics, such as color and texture, to readily discernible processes and components of the soil. Soil, the "underfoot landscape", is best understood by relating it to common experience.

6. Understand the basic ways by which humans have affected soils in world regions; these include soil salinization and accelerated soil erosion.

KEY TERMS AND CONCEPTS

Provide a short definition or description of the following key words and concepts from the chapter.

Soil Basics

Structure

Peds

Blocky

Prismatic (Columnar)

Granular

Crumb

Texture

Sand

Silt

Clay

Colloids

Loam

Color

Iron Oxides

Humus

Calcium Carbonate

Quartz/Sandy Soils

Processes

Translocation

Leaching

Duricrust

Components

    Inorganic Mineral Material

    Organic Matter

        Humus

        Clay-Humus Complex

        Cation Exchange Capacity (CEC)

Pore Spaces

    Soil Gases

    Soil Water

        Field Capacity

        Capillary Tension

        Wilting Point

        Acidity

        Alkalinity

Soil Horizons

    Eluviation

    Illuviation

    Soil Profile

    O Horizon

    A Horizons

    B Horizons

    C Horizons

    Topsoil

    Subsoil

    Solum

<u>Soil Development Factors</u>

     Climate

     Organisms

     Relief

     Parent Material

     Time

<u>Climate</u>

     Precipitation

     Temperature

     Vegetation Type

     Wetness and Dryness

     Microscopic Organisms

<u>Organisms</u>

     Bacteria

     Nitrogen Fixation

     Legumes

     Protozoa

     Worms

     Insects

     Trees, Other Macroscopic Vegetation

<u>Relief</u>

     Aspect

     Slope

     Vegetation

Parent Material

  Elements of the Crust

    Oxygen

    Silicon

    Aluminum

    Iron

    Calcium, Sodium, Potassium, Magnesium

  Regolith

  Residual Materials

  Transported Materials

Time

  Climate Influences

  Rate of Formation

Classification

  Physical Characteristics

  Biologic Characteristics

  Chemical Characteristics

  Spatial Variation

  Human Modification

7th Approximation

  Present Characteristics (Descriptive)

  Includes Modified Soils

  Descriptive Names

  Hierarchy of Levels

The Ten Soil Orders

Aridisol

Mollisol

Spodosol

Alfisol

Ultisol

Oxisol

Vertisol

Entisol

Inceptisol

Histosol

TESTING: TRUE/FALSE

_____ 1. The percentage of a typical soil that is solid (by volume is typically 50%.

_____ 2. A particle less than 0.002 millimeters in diameter is a clay particle consisting of a clay mineral.

_____ 3. The leading cause of soil loss throughout the world is erosion.

_____ 4. The number of soil orders included in the 7th
Approximation is eight.

_____ 5. A granular soil structure might be found in a
prairie soil having a high humus content.

_____ 6. The loose material on the surface of the moon
cannot be considered to be a soil because it has no
organic matter.

_____ 7. On a soil texture triangle the relative proportions
of sand, gravel, and clay are plotted.

_____ 8. An entisol is a recent, poorly developed soil with
little or no profile development.

_____ 9. Spodosols are found in coniferous forests because
forest fires provide the ash component that is
translocated to the lower part of the A horizon.

_____ 10. The removal of solid or dissolved materials from
one horizon to another is called eluviation.

TESTING: MATCHING

11. _____ red color   A. presence of organic matter

12. _____ O horizon   B. histosols

13. _____ white nodules   C. important soil organism

14. _____ bacteria   D. accumulation of organic
                                  debris
15. _____ bog soils
                               E. found in dry climates
16. _____ duricrust
                               F. calcium carbonate particles
17. _____ parent material
                               G. solidified soil material
18. _____ infertile soil
                               H. underlying bedrock
19. _____ aridisols
                               I. presence of iron oxides
20. _____ black color
                               J. acid soil

TESTING: MULTIPLE CHOICE

Choose the best response for each of the following multiple
choice questions. Each question has only one correct answer.

21. Which of the following is not included in determination
     of soil texture?
     a. sand
     b. gravel
     c. silt
     d. clay

22. A mature soil profile provides a good example of a system
     in _____ with its environment.
     a. dynamic equilibrium
     b. steady state
     c. retarded development
     d. dynamic instability

23. Which of the following is not a characteristic of a
     spodosol?
     a. low fertility
     b. high humus in the O horizon
     c. associated with needleleaf forests
     d. alkaline soils

24. The deposition of translocated soil materials is given
     the special name:
     a. eluviation
     b. illuviation
     c. precipitation
     d. morphogenesis

25. The soil particles which are so small that they can
     remain suspended indefinitely in water are called:
     a. colloids
     b. cations
     c. anions
     d. clay-humus complex

26. The maps accumulated which describe the soil types and
     distribution within a county are available from the:
     a. Department of Forestry
     b. Geological Survey
     c. Soil Conservation Service
     d. Department of Agriculture

27. The number of soil series recognized in the United
     States is closest to:
     a.   20,000
     b.   14,000
     c.   10,000
     d.   5,000

28. Most soil color is imparted by the following two
     substances:
     a.   iron and calcium
     b.   iron and inorganic minerals
     c.   organic matter and soil water
     d.   iron and humus

29. The 7th Approximation:
     a.   is a 19th century soil classification
     b.   is a descriptive system of classification
     c.   is a genetic system of classification
     d.   has names which are arbitrary

30. The texture of a soil is of practical significance in:
     a.   the workability and color
     b.   the ability to hold water and support crops
     c.   the ability to hold water and the workability
     d.   the ability to support crops and resist erosion

31. Which of the following characteristics is not related to
     a soil's structure?
     a.   ped shape
     b.   ped texture
     c.   ped organization
     d.   ped size

32. In general, the greatest activity of soil microscopic
     organisms and the least accumulation or organic debris
     is in:
     a.   the tropics
     b.   the deserts
     c.   the middle latitudes
     d.   the high latitudes

33. A possible solution to the salinization of soils is:
     a.   reducing the amount of water placed on crops
     b.   restrict drainage so that salts cannot accumulate
     c.   install tile drains
     d.   use irrigation water containing a higher salt content

34.   The ability of a soil to hold nutrient cations is called:
      a.   the cation exchange capacity
      b.   the wilting fertility
      c.   the cation/anion ratio
      d.   the soil buffering capacity

35.   Under favorable circumstances, the number of earthworms
      inhabiting a soil:
      a.   can exceed 1,000,000 per acre
      b.   can weigh over 1,000 pounds per acre
      c.   can ingest a tremendous amount of soil
      d.   each of the above answer is correct

TESTING: SHORT ANSWER AND ESSAY

1.   How rapidly does soil evolve?   Under what conditions is
     this evolution most rapid?

2.   What are the principal soil orders in the United States
     which support our most productive agriculture?   Where are
     they located?

3. Make a list of the principal chemical and physical
    characteristics of ocean water. What is the origin of
    these characteristics?

4. How is soil texture related to the field capacity and to
    the wilting point of a soil?

5. Make a list of common soil colors and the materials that
    are responsible for the color? Is there any relationship
    between soil color and fertility?

6.   Describe the texture of a clayey silty loam.

7.   Describe the texture of a clayey silty loam.

8. In what ways can American agricultural practices be
    traced to accelerated soil erosion? Can science
    always be trusted to solve the ongoing problem of
    soil deterioration?

PLACE AND PHYSICAL FEATURE LOCATION

SOUTHERN AFRICA MAP EXERCISE

Locate the following by name or number on the map of **Southern Africa**. (NOTE: you may wish to sketch in the major landforms).

COUNTRIES

1. Zaire
2. Lesotho
3. Rwanda
4. Malawi
5. Congo
6. Swaziland
7. Zimbabwe
8. Botswana
9. Uganda
10. Gabon
11. Tanzania
12. Zambia
13. Namibia
14. Malagasy Republic
15. South Africa
16. Angola (and Cabinda)
17. Mozambique
18. Kenya
19. Burundi

MAJOR CITIES

33. Durban
34. Nairobi
35. Cape Town
36. Lusaka
37. Kampala
38. Dar es Salaam
39. Johannesburg
40. Luanda
41. Kinshasa
42. Pretoria
43. Mombasa
44. Lilongwe
45. Windhock
46. Maputo
47. Salisbury
48. Zanzibar

PHYSICAL FEATURES

20. Namib Desert
21. Orange River
22. Lake Victoria
23. Congo (Zaire River)
24. Cape of Good Hope
25. Zambezi River
26. Kalahari Desert
27. Lake Nyasa
28. Lake Tanganyika
29. Limpopo River
30. Mozambique Channel
31. Drakensberg Mountains
32. Madagascar

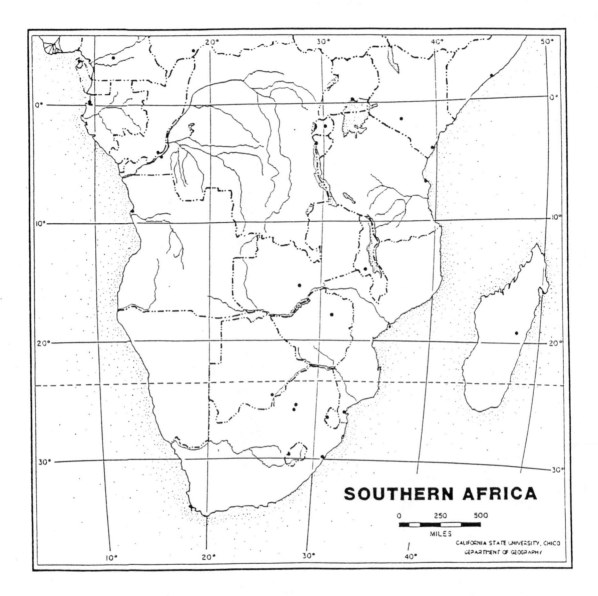

SOUTHERN AFRICA

0   250   500
MILES

CALIFORNIA STATE UNIVERSITY, CHICO
DEPARTMENT OF GEOGRAPHY

C H A P T E R   13

INTRODUCTION TO LANDFORMS

1. Introduce the conceptual basis for studying geomorphology and landforms.

2. Describe the basic forces that operate on a landscape to create the features we see as landforms. Discuss the energy sources for these processes.

3. Briefly look at the classification systems for landforms. In general, it is widely acknowledged that no rigorous classification system exists.

4. present the nomenclature of landforms. This discussion is not descriptive, but instead gives examples of some types of extensive landscapes.

5. Review the geologic timetable and develop a temporal framework for discussions of landforms.

6. Examine the location of some of the major cordilleran belts of the world and discuss their ages. The appearance and relief of many mountain ranges can be related to their age.

7. Grasp the basis for a genetic approach to the study of the earth's landforms.

KEY TERMS AND CONCEPTS

Provide a short definition or description of the following key words and concepts from the chapter.

General concepts

    Topography

    Geomorphology

        Geographical

        Geological

    Structure

    Process

    Time

Geologic Time

    Quaternary

        Recent

        Pleistocene

    Tertiary

    Mesozoic

    Paleozoic

    Precambrian

Relict Landforms

Energy Sources

    Solar

    Internal

    Gravity

Theories

    Catastrophism

    Uniformitarianism

Classification

    Relief

    Plains

        Interior Plains

        Coastal Plains

        Plains With Areas of High Relief (hills and mountains)

    Plateau

        Plateaus With Canyons or Marginal Escarpments

    Canyon (gorge)

Global Distribution

    Continents

    Cordilleran Belts

Genetic Approaches to Landform Study

    Tectonic Forces

    Diastrophism

        Folding

        Faulting

    Vulcanism

        Intrusive

        Extrusive

Gradational Forces

     Water, Ice, and Air

          Flowing Water

          Ice Expansion

          Glaciers

          Wind

          Ocean Waves and Current

          Gravity

     Weathering

     Erosion

     Deposition

<u>TESTING: TRUE-FALSE</u>

_____   1. Basic misconceptions regarding the age of the
              earth can be related to literal interpretations
              of the Bible.

_____   2. The relict glacial features covering much of North
              America were produced during the Tertiary.

_____   3. The Grand Canyon, an excellent example of a steep
              gorge, is cut ten miles below the Colorado
              plateau.

_____   4. The three substances accomplishing most of the
              erosion of the surface are ice, air, and water.

_____   5. The major landform producing forces are
              gradational and tectonic.

_____   6. Tectonic forces are powered by the decay of
              radioactive elements.

_____   7. Gravity has resulted in the general leveling of
              the earth's surface into a continuous plain.

_____   8. The principle of Uniformitariansim was first
              proposed by William M. Davis.

ſ

_____ 9.   Older and lower mountain systems are often located
              on the opposite sides of continents from the
              cordilleran belts.

_____ 10. Plateaus are created by the folding of sediments
              in newly formed mountain ranges.

TESTING: MULTIPLE CHOICE

Choose the best response for each of the following multiple
choice question. Each question has only one correct answer.

11. The number of large continental blocks is:
     a.   4
     b.   6
     c.   8
     d.   10

12. The branch of earth science devoted to the systematic
     study of earth landforms is called:
     a.   geomorphology
     b.   geology
     c.   landscape architecture
     d.   planimetric analysis

13. Which continental mass is not partially located in the
     Southern Hemisphere?
     a.   Australia
     b.   Antarctica
     c.   Africa
     d.   Eurasia

14. The belief that past cataclysmic events have produced the
     major features of the earth is called:
     a.   uniformitarianism
     b.   steady state
     c.   dynamic equilibrium
     d.   catastrophism

15. One reason why the earth's landforms may be unique in the
     Solar System:
     a.   is the composition of our atmosphere
     b.   is the composition of the bedrock
     c.   is the presence of nitrogen in the atmosphere
     d.   is our close distance to the sun

16. Gradational forces:
     a.   build mountains
     b.   raise the surface and increase relief
     c.   are powered by the sun and by gravity
     d.   are less powerful than erosive forces

17. A plateau resulting from the extrusion of basaltic lava
     is:
     a.   the Colorado Plateau
     b.   the Columbia Plateau
     c.   the Anatolian Plateau
     d.   the High Antarctic Plateau

18. Which of the following is not a tectonic force?
     a.   folding
     b.   faulting
     c.   extrusive vulcanism
     d.   chemical weathering

19. The geologic period which was associated with the
     development of flowering plants is the:
     a.   Pleistocene
     b.   Cretaceous
     c.   Permian
     d.   Ordovician

20. The continent, excluding Antarctica, which has the
     greatest proportion of ice caps is:
     a.   South America
     b.   Africa
     c.   Eurasia
     d.   North America

TESTING: SHORT ANSWER AND ESSAY

1. Explain the differences between catastrophism and
    uniformitarianism. Which principle is regarded as
    best representing the landscapes developing during
    the Pleistocene?

2. Briefly describe the energy sources for gradational
    processes and for tectonic processes on the earth's
    surface. Why do these differ?

3. In what important ways do the geomorphologic processes
    of the earth differ from those of other planets?

4.   Describe the global distribution of major mountain belts.

5.   Make a list of the landforms in the area immediately
     around your home town. Do the words capture a unique
     landscape? What is a basic problem of the nomenclature
     of geomorphology?

6.   Briefly describe each of the more major descriptive
     classes of landforms. Give examples from your home
     State, if possible, for each type.

PLACE AND PHYSICAL FEATURE LOCATION

NORTHERN AFRICA MAP EXERCISE

Locate the following by name or number on the map of **Northern Africa**. (NOTE: you may wish to sketch in the major landforms).

COUNTRIES

1. Morocco
2. Ethiopia
3. Guinea-Bissau
4. Kenya
5. Mali
6. Ivory Coast
7. Central African Republic
8. Tunisia
9. Ghana
10. Sudan
11. Uganda
12. Congo
13. Libya
14. Togo
15. Niger
16. Egypt
17. Senegal
18. Benin
19. Gabon
20. Sierra Leone
21. Somalia
22. Cameroon
23. Mauritania
24. Liberia
25. Algeria
26. Chad
27. Guinea
28. Nigeria
29. Burkina Faso
30. Gambia

PHYSICAL FEATURES

31. Madeira Islands
32. Atlas Mountains
33. Sahara
34. Lake Chad
35. Niger River
36. Nile River
37. Zaire River
38. Lake Victoria
39. Strait of Gibraltar
40. Canary Islands
41. Arabian Desert
42. Red Sea
43. Gulf of Suez
44. Gulf of Guinea

MAJOR CITIES

45. Abidjan
46. Addis Ababa
47. Cairo
48. Djibouti
49. Algiers
50. Tunis
51. Accra
52. Kampala
53. Dakar
54. Casablanca
55. Lagos
56. Khartoum
57. Nairobi
58. Rabat

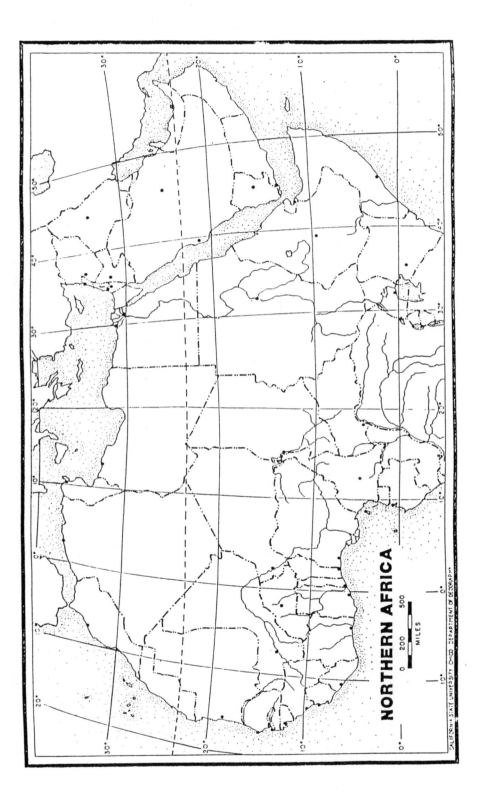

NORTHERN AFRICA

0   200   500
MILES

CALIFORNIA STATE UNIVERSITY CHICO DEPARTMENT OF GEOGRAPHY

C H A P T E R   14

EARTH'S DYNAMIC LITHOSPHERE

LEARNING OBJECTIVES

1. Review the basic structure of the earth's interior and discriminate between the various layers -- the <u>crust</u>, the <u>mantle</u>, and the <u>core</u>.

2. Be able to name the basic classes of igneous rocks and the main mineral constituents of intrusive rocks.

3. Examine the processes which operate to form the sedimentary rocks and discriminate between the three basic types -- clastic, organic, and chemical sedimentary rocks.

4. Study the processes of metamorphism that result in the major types of metamorphic rocks.

5. Develop an awareness of the slow development of evidence in support of the plate tectonic/continental drift theory.

6. Discover how rock materials are continually cycled through the earth and the important role of plate tectonics in this process.

7. Investigate the sources of energy for mountain building, and the particularly important role of the decay of the radioactive elements uranium and thorium.

8. Learn the three basic types of motions which lithospheric plates can make; relate these motions to the occurrences of island arcs, volcanos, and earthquakes.

9. Explain why the sediments in the ocean are so young, and why they are thinnest near the center of ocean basins.

10. Relate the processes of plate tectonics and continental drift to earthquakes and volcanism in our world today.

11. Become aware of the complex internal structure of the earth, the names of the layers, their behavior, and the role of seismology in gaining information about these zones.

12. Provide an explanation of the absence of mountains in shield areas of the earth, and their frequent presence along coastal boundaries of North and South America.

13. List the early and modern (post-WWII) evidence for plate tectonics. Discuss how major revolutions in scientific thought transpire.

14. Understand the location of world regions where geothermal heat is present at the earth's surface -- and the manner by which it can be used as an energy source.

15. Provide an explanation for the Himalaya Mountains.

16. Discuss the geologic mechanisms behind one of the great wonders of the United States: Yellowstone National Park.

KEY TERMS AND CONCEPTS

Provide a short definition or description of the following key words and concepts from the chapter.

Structure of the Earth

>   Inner Core
>
>   Outer Core
>
>   Mantle
>
>   Mohorovicic Discontinuity
>
>   Crust
>
>   Aesthenosphere
>
>   Lithosphere
>
>   Oceanic Crust

Sima

Continental Crust

Sial

Rocks and Minerals

Rock

Mineral

Silicate Minerals

Sediments

Strata

Bedding Planes

Intrusive Igneous Rocks

Granite

Diorite

Gabbro

Pluton

Batholith

Extrusive Igneous Rocks

Basalt

Rhyolite

Obsidian

Pumice

Sedimentary Rocks

Clastic Sedimentary Rocks

Conglomerate

Sandstone

                        Siltstone

                        Shale

                Organic Sedimentary Rocks

                        Limestone

                        Coal

                Chemical Sedimentary Rocks

                        Limestone

                        Dolomite

                        Chert

                Evaporite Sedimentary Rocks

                        Halite

                        Gypsum

                Salt Domes

Metamorphic Rocks

        Regional metamorphism

        Contact Metamorphism

                Quartzite

                Slate

                Schist

                Gneiss

                Marble

                Anthracite Coal

Rock Cycle

Plate Tectonics

        Lithospheric Plates

        Plate Tectonics

Sea-Floor Spreading

Subduction

Suturing

Isostatic Uplift

Continental Drift

Rift Valley

Submarine Trenches

Island Arcs

Terranes

Shields

Evidence for Plate Tectonics (relate the following):

Seismographic Records

Fit on Continents

Matching Structures

Matching Fossils

Matching Flora and Fauna

Paleomagnetic Evidence

Midoceanic Mountain Ranges (Ridges)

Age of Deep Ocean Sediments

Thickness of Deep Ocean Sediments

Radioactive Energy Sources

TESTING: TRUE/FALSE

_____   1.   The earth's core occupies eighty percent of the
              planet's total volume.

_____   2.   The asthenosphere is located between the mantle
              and the crust of the earth.

_____   3.   Plutons are comprised of intrusive igneous rocks
                such as granite.

_____   4.   Plate tectonics became an accepted part of
                geologic theory at the turn of the 20th century.

_____   5.   The instrument which records shock waves in the
                earth is a seismograph.

_____   6.   The metamorphic equivalent of bituminous coal is
                called anthracite.

_____   7.   Magma which makes its way to the earth's surface
                may be extruded as lava or thrown out as a
                batholith.

_____   8.   Sea-floor spreading and subduction are both
                <u>convergent</u> plate movements.

_____   9.   Suturing refers to the fusion of two plates into
                one single larger plate.

_____   10.   The oldest known ocean rock is only 180 million
                years old, while the earth is 5 billion years old.

_____   11.   A deep, steep-sided trough which runs the length
                of the continent of Africa is an example of a rift
                valley.

_____   12.   A sedimentary rock resulting from the evaporation
                of ocean water is halite.

_____   13.   The development of a layered interior to the earth
                developed near the end of geologic time, just
                prior to the Pleistocene.

_____   14.   The mantle, core, and crust can be distinguished
                on the basis of their physical state.

TESTING: MATCHING

Letters may be used more than once:

A.   EXTRUSIVE IGNEOUS ROCK   _____ 15.   Sandstone

B.   INTRUSIVE IGNEOUS ROCK   _____ 16.   Gneiss

C.   CLASTIC SEDIMENTARY ROCK   _____ 17.   Basalt

D.   ORGANIC SEDIMENTARY ROCK   _____ 18.   Coal

E.   CHEMICAL SEDIMENTARY ROCK   _____ 19.   Conglomerate

F.   METAMORPHIC ROCK   _____ 20.   Shale

                                     _____ 21.   Marble

                                     _____ 22.   Limestone

                                     _____ 23.   Granite

                                     _____ 24.   Halite

TESTING: MULTIPLE CHOICE

25.   The four most abundant elements in the crust of the earth
      include:
      a.   iron, aluminum, potassium, and nickel
      b.   oxygen, silicon, aluminum, and iron
      c.   oxygen, calcium, aluminum, and iron
      d.   iron, oxygen, sodium, and silicon

26.   The least abundant of the three main rock types are:
      a.   igneous rocks
      b.   metamorphic rocks
      c.   sedimentary rocks

27.   Specific gravity is commonly given in units of grams per
      cubic centimeter. The specific gravity of the oceanic
      crust is:
      a.   twice that of water
      b.   three times that of water
      c.   five times that of water
      d.   eight times that of water

28.   Located along the midoceanic ridges are:
      a.   active submarine volcanos
      b.   old oceanic sediments
      c.   outcrops of granitic rock
      d.   evaporite deposits

29. The metamorphic equivalent of granite or diorite is:
    a. shale
    b. phyllite
    c. gneiss
    d. schist

30. The igneous intrusive rocks diorite and gabbro
    a. each contain substantial amounts of quartz and
       feldspars
    b. are characteristic of extrusive lavas
    c. contain substantial quantities of minerals rich in
       iron and magnesium
    d. are the extrusive equivalent of basalt

31. The mantle of the earth:
    a. contains four-fifths of the earth's volume
    b. is composed of rocks having a granitic composition
    c. cannot transmit seismic waves
    d. is brittle

32. An igneous extrusive rock called obsidian is used as the
    raw material for arrowheads because:
    a. it is toxic
    b. it cooled so rapidly that it is glassy
    c. it breaks easily along bedding planes
    d. it is a common rock type

33. The basic unit into which the lithosphere is broken is a:
    a. plate
    b. subduction zone
    c. suture zone
    d. rift valley

34. The distribution of continental tectonic features is most
    influenced by:
    a. spreading centers
    b. oceanic ridges
    c. subduction zones
    d. convergent plate boundaries

35. Large granitic plutons are called:
    a. pumice
    b. batholiths
    c. rift zones
    d. island arcs

36. The energy processes responsible for plate tectonics are
    driven by the radioactive decay of:
    a. uranium and thorium
    b. plutonium and strontium
    c. plutonium and uranium
    d. thorium and plutonium

37. What region of the western United States is an area
     situated directly above a hot spot (similar to Hawaii)?
     a.   the Geysers in California
     b.   the Black Hills
     c.   Mount Saint Helens
     d.   Yellowstone Park

38. Particularly stable crustal areas are called:
     a.   island arcs
     b.   shields
     c.   batholiths
     d.   monoliths

39. The magnetic record of oceanic volcanics:
     a.   documents reversals in the magnetic field of the
          earth
     b.   implies that plate tectonics is occurring
     c.   could not be read until after World War II
     d.   each of the above answers is correct

40. The number of large major plates is:
     a.   4
     b.   6
     c.   8
     d.   10

41. The movements of the lithospheric plates are caused by:
     a.   gravity
     b.   the gradational forces
     c.   convection currents in the mantle
     d.   magnetic activity in the core

42. Which continental mass is not partially located in the
     Southern Hemisphere?
     a.   Australia
     b.   Antarctica
     c.   Africa
     d.   Eurasia

TESTING: SHORT ANSWER AND ESSAY

1. Name the three basic categories of rocks. What type of
    rocks are found in the middle oceanic ridges? What type
    of rocks are found in the ocean basins? What type is
    most common at the continental surface?

2. Describe the rock cycle, giving an example of what
    happens to granitic material originally intruded above
    a subduction zone.

3. Differentiate between the <u>intrusive</u> and the <u>extrusive</u>
    igneous rocks, and provide examples of each.

4. The lithospheric plates are driven by convection
    currents; explain how this is possible.

5. Name the principal differences between oceanic crust and
    continental crust.

6. Describe a situation in which halite, a readily soluble
    mineral (common table salt is halite) might develop.

7. What are the main lines of evidence for continental
    drift/plate tectonics? Provide a chronology for the
    accumulation and acceptance of this evidence within this
    century.

8. Explain the three main types of tectonic plate movement
    and provide a good example of each.

9. What is the correlation over the earth between
    lithospheric plate boundaries and the following: island
    arcs, earthquake zones, volcanic activity, shields?

PLACE AND PHYSICAL FEATURE LOCATION

AUSTRALIA AND NEW ZEALAND MAP EXERCISE

Locate the following by name or number on the map of **Australia and New Zealand**. (NOTE: you may wish to sketch in the major landforms).

COUNTRIES

1. New Caledonia
2. Vanuatu
3. Solomon Islands
4. New Zealand
5. Tonga
6. Fiji
7. Western Samoa
8. Papua New Guinea

AUSTRALIAN STATES:

9. New South Wales
10. Queensland
11. South Australia
12. Victoria
13. Western Australia
14. Northern Territory
15. Tasmania

PHYSICAL FEATURES

26. Torres Strait
27. Bass Strait
28. Timor Sea
29. Great Dividing Range
30. Great Barrier Reef
31. Great Australian Bight
32. Gulf of Carpentaria
33. Cook Strait
34. Arafura Sea
35. Coral Sea
36. Lake Eyre
37. Darling River
38. Murray River
39. Nullarbor Plain
40. Gibson and Great Sandy Deserts
41. North Island
42. South Island

MAJOR CITIES

16. Perth
17. Canberra, A.C.T. (Australia Capital Territory)
18. Sydney
19. Wellington
20. Auckland
21. Brisbane
22. Melbourne
23. Adelaide
24. Alice Springs
25. Darwin

Earth's Dynamic Lithosphere

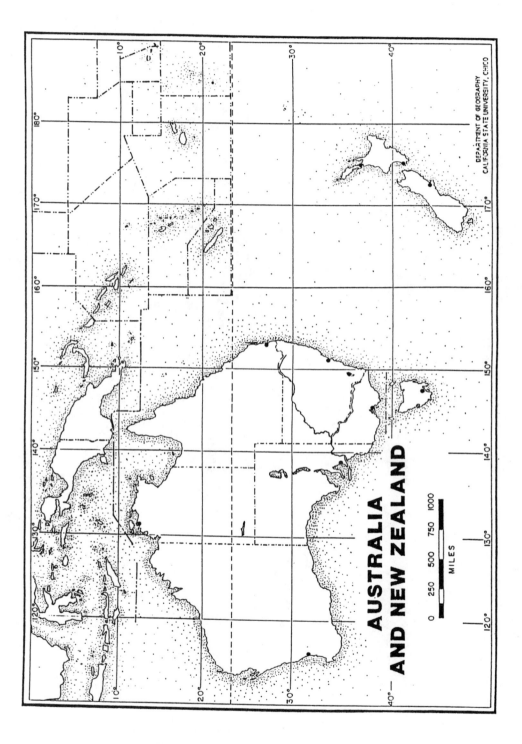

AUSTRALIA AND NEW ZEALAND

0   250   500   750   1000
MILES

DEPARTMENT OF GEOGRAPHY
CALIFORNIA STATE UNIVERSITY, CHICO

C H A P T E R   15

DIASTROPHIC AND VOLCANIC LANDFORMS

LEARNING OBJECTIVES

1. Describe the tectonic forces which operate to cause crustal rocks to undergo folding.

2. Identify the types of structural landforms which are created by folding.

3. Discuss the consequences of erosion upon the structural ridges and valleys in a folded region.

4. Compared the structures of domes and basins with regions of linear folded topography and provide examples of each.

5. Describe the forces which operate to produce faulting movements as a response of rocks to tectonic stress.

6. Differentiate between the four major kinds of faults in terms of direction of rock movement and resultant structural landform features.

7. Discuss the relationship between major planetary fault zones, such as the San Andreas Fault of California, and the concept of plate tectonics.

8. Identify the faulting forces which result in grabens and horsts and relate these to both the basin and range topography of the United States and the East African Rift Valley.

9. Outline the causes of earthquakes and be able to describe how earthquakes are both measured and related to one another using a scale of earthquake magnitude.

10. Describe the types of hazards that earthquakes pose to human activity and settlement.

11. Identify the main processes of extrusive volcanism in terms of the three main types of volcanic mountains and the extrusive igneous materials involved.

12. Describe the basic global distribution of active volcanic activity and identify historic and present-day distributions of fissure eruptions.

13. Identify the types of volcanic hazards and associated damaging events which should be considered by anyone living in an area of active volcanism.

14. Discuss intrusive volcanism in terms of the rock materials involved, the types of plutonic features, and the resultant impacts on the topography of the earth's surface.

KEY TERMS AND CONCEPTS

Diastrophism

Anticline

Syncline

Recumbent Fold

Differential Erosion

Reversal of Topography

    Anticlinal Valleys

    Synclinal Ridges

Homoclinal Ridges and Valleys

Dome

Basin

Faulting

    Fault Line

    Fault Plane

    Fault Scarp

Normal Fault

Reverse Fault

Transcurrent (Strike-Slip) Fault

Thrust Fault

Horst and Graben

Rift Valleys

Basin and Range Topography

Earthquake

Seismograph

P Waves and S Waves

Richter Scale

Tsunami

Extrusive Volcanism

Volcano

Pacific "Ring of Fire"

Pyroclastic Material

Cinder Cone

Shield Volcano

Composite Volcano

Caldera

Nuees Ardentes

    Lava Flow

        Flood Basalts

        Fissure Eruptions

        Lava Plateau

Intrusive Volcanism

    Batholith

    Laccolith

    Sill

    Dike (Dike Ridge)

    Volcanic Neck

TESTING: TRUE-FALSE

_____   1. Volcanic eruptions are a classical illustration of
              the forces known as diastrophism.

_____   2. Most folding takes place in response to lateral
              compressional forces operating upon sedimentary
              rocks.

_____   3. An upfolded area is called an anticline whereas a
              downfolded area is termed a syncline.

_____   4. In the initial structural topography of a region,
              the anticline appears as a valley while the
              syncline is a ridge.

_____   5. The two-dimensional surface of a fault as it
              extends into the ground is termed the fault line.

_____   6. A steep fault scarp is often associated with an
              area of normal faulting.

_____   7. The forces which create reverse and thrust faults
              act in direct opposition to one another.

_____   8. The San Andreas Fault of California is a classic
              example of a normal fault exposed at the earth's
              surface.

_____   9.   The San Andreas Fault of California is in actuality a tectonic plate boundary.

_____   10.   A horst is a raised block bounded by two normal faults.

_____   11.   Earthquake activity produces shaking which acts to break rock apart thereby creating fault lines.

_____   12.   The Richter Scale is a logarithmic scale which relates the magnitude of earthquakes to one another.

_____   13.   A cinder cone is a small volcano composed wholly of pyroclastic material.

_____   14.   Calderas are formed in roughly the same manner as domes and basins.

_____   15.   Batholiths have formed as deep-seated plutonic emplacements and therefore are never exposed at the earth's surface.

_____   16.   Shiprock New Mexico is a classic example of a volcanic neck.

_____   17.   A tsunami is a sea wave which is generated by tectonically induced movements of the sea floor.

TESTING: MATCHING

A. ANTICLINE

B. GRABEN

C. BASIN

D. NORMAL FAULT

E. CINDER CONE

F. SYNCLINE

G. TRANSCURRENT FAULT

H. SHIELD VOLCANO

I. DOME

J. COMPOSITE VOLCANO

K. FLOOD BASALT

L. LACCOLITH

M. BATHOLITH

N. CALDERA

O. THRUST FAULT

P. DIKE

Q. SILL

R. TSUNAMI

S. SEISMOGRAPH

T. NUEES ARDENTES

____ 18. The smallest type of volcano

____ 19. The largest plutonic rock feature

____ 20. A seismic sea wave

____ 21. A rounded anticlinal structure

____ 22. An instrument which records earthquake waves

____ 23. Superheated volcanic gas

____ 24. An upfolded ridge

____ 25. Broad volcano composed of basaltic rock

____ 26. The most common vertical fault type

____ 27. A fault caused by extreme compressional forces

____ 28. Sheets of magma injected between sedimentary rock layers

____ 29. A horizontal fault type

____ 30. Small, dome-shaped plutonic intrusion

____ 31. A volcano associated with explosive eruptions

____ 32. A downfolded valley

____ 33. A downdropped area between two normal faults

TESTING: MULTIPLE CHOICE

34.   A landform feature that results from diastrophism:
      a.   batholith
      b.   anticline
      c.   cinder cone
      d.   flood basalt

35.   The overturning of rock strata in the fold flanks
      produces a:
      a.   syncline
      b.   dome
      c.   basin
      d.   recumbent fold

36.   Following a reversal of topography an original anticline
      will become a:
      a.   homoclinal fold
      b.   dome
      c.   ridge
      d.   valley

37.   The Black Hills of South Dakota is a classic example of
      a:
      a.   dome
      b.   basin
      c.   synclinal valley
      d.   nappe

38.   A fault with a predominantly horizontal slippage is a:
      a.   normal fault
      b.   reverse fault
      c.   transcurrent fault
      d.   thrust fault

39.   A high-angle vertical fault displacement can produce a:
      a.   fault scarp
      b.   syncline
      c.   dome structure
      d.   nappe

40.   Which of the following is not produced by graben
      faulting:
      a.   Death Valley, California
      b.   the Appalachians
      c.   the Basin and Range of the Western United States
      d.   the Rift Valley System of East Africa

41. Which of the following would <u>not</u> be associated with the Pacific "Ring of Fire":
    a. volcanic activity
    b. earthquake
    c. tsunamis
    d. synclines

42. Which of the following is <u>not</u> associated with extrusive volcanic activity:
    a. batholith
    b. cinder cone
    c. fissure flow
    d. nuees Ardentes

43. A volcanic mountain composed of layers of lava and pyroclastic material is a:
    a. cinder cone
    b. shield volcano
    c. composite volcano
    d. plateau basalt

44. A classic world region of basaltic lava flows and shield volcanos:
    a. Japan
    b. Hawaiian Islands
    c. the Cascades Mountains
    d. the Appalachians

45. The most violent and explosive volcanic eruptions are produced by:
    a. composite volcanos
    b. cinder cones
    c. shield volcanos
    d. fissure eruptions

46. Crater Lake, Oregon is a classic example of:
    a. a fissure flow
    b. a caldera
    c. a cinder cone
    d. a volcanic neck

47. Which of the following does <u>not</u> relate to batholithic emplacements:
    a. is exposed in the cores of many mountain ranges
    b. is formed by intrusive volcanism
    c. is composed of pyroclastic debris
    d. contains valuable metal deposits

48. Laccoliths are:
    a. larger than batholiths
    b. smaller than sills or dikes
    c. associated with plutonic basins
    d. small dome-shaped plutonic emplacements

49. Sills differ from dikes in that:
    a. dikes are injected vertically across rock strata
    b. sills are volcanic in origin
    c. dikes are flat and sills are wall-like
    d. sills are pyroclastic while dikes are not

50. The 1883 eruption of Krakatau:
    a. produced a large tsunami
    b. did not produce a caldera
    c. had 95 percent of the erupted material from the
       island
    d. all of the above

51. The Loma Prieta Earthquake of October 17, 1989:
    a. was on the San Andreas Fault
    b. had its epicenter in the Santa Cruz Mountains
    c. had a Richter magnitude of 7.1
    d. all of the above

TESTING: SHORT ANSWER AND ESSAY

1.   What types of structural landforms might be found in an
     area which has undergone extensive folding activity?
     What happens to these structures after erosion has worked
     upon them?

2.   How are domes and basins formed? What types of landforms
     are formed? Provide good examples of each.

3.   What are the four main types of fault in terms of
     direction of faulting movement? What kinds of resultant
     structural features are created?

4.   What created the San Andreas Fault of California and what
     type of fault is it? In what ways does this fault zone
     illustrate the earthquake hazard which exists in such
     tectonically active areas? What major seismic event
     occurred in 1989 along this fault zone?

5. What are earthquakes? How are the measured? What world
    regions seem particularly prone to earthquake activity?

6. Compare and contrast the three main types of volcanic
    mountains: cinder cones, shield volcanos, and composite
    volcanos. In what ways are each of these formed and what
    kinds of volcanic hazards are posed by each to humans?

7. What is the basic global distribution of volcanic
    activity? What is the significance of the Pacific "Ring
    of Fire" in this regard? Are there any relationships
    between this pattern and the pattern for earthquake
    activity noted earlier?

8.   What types of landform features result from intrusive
     volcanism? What are the resultant impacts on the
     topography of the earth's surface?

9.   What are tsunamis and where do they occur? Why could
     they be considered to be damaging events?

C H A P T E R   16

WEATHERING, MASS WASTING, AND KARST TOPOGRAPHY

LEARNING OBJECTIVES

1. Discuss the importance of weathering in preparing surface rock materials for the processes of gradation.

2. Compare and contrast physical and chemical weathering in terms of their interrelated impacts on surface rock materials.

3. Outline and explain the main processes and resultant landscape features of physical weathering.

4. Outline and explain the main reactions and consequent impacts upon rock materials of the main types of chemical weathering.

5. Explain the basic global distributions of chemical and physical weathering processes as these differ with regard to climatic and lithologic conditions under which they operate.

6. Explain the importance of gravity in providing the impetus for the gradational process of mass wasting.

7. Discuss the significance of the "angle of repose" in mass wasting.

8. Identify the way by which human activities have affected slope angles and have thereby initiated mass movement.

9. Compare and contrast the major forms of mass wasting of slope materials and identify the resultant landform features.

10. Describe the specific environmental conditions which favor the development of karst topography and identify the main world regions affected by solution weathering.

11. Differentiate between the landforms associated with an area of karst topography and those found in areas affected by usual stream activities.

12. Identify and explain the specific environmental hazards to humans occupying an area which is being affected by active solution weathering.

KEY TERMS AND CONCEPTS

Weathering

Physical Weathering Processes and Features

Unloading

Exfoliation Sheets and Domes

Frost Wedging

Boulder Fields

Talus Cones

Thermal Expansion and Contraction

Salt Crystal Growth

Biological Forces

Chemical Weathering Process and Features

Hydration

Hydroloysis

Oxidation

Carbonation

Solution

Mass Wasting

Angle of Repose

<u>Types of Mass Movements</u>

> Avalanche
>
> Rockslide
>
> Debris Slide
>
> Mudflow
>
> Slump
>
> Earth Flow
>
> Creep
>
> Debris Avalanche

<u>Karst Topography</u>

> Sinkhole
>
> Karst Plain
>
> Uvala
>
> Sinking Creek
>
> Cave
>
> Travertine Features (Stalactites and Stalagmites)
>
> Haystack Hill
>
> Tower Karst

TESTING: TRUE-FALSE

_____   1.   Weathering may be defined as the combined action
               of physical and chemical processes that breakdown
               bedrock.

_____   2.   Unloading can lead directly to the creation of
               exfoliation sheets and ultimately to exfoliation
               domes.

_____ 3. Frost wedging is a textbook example of chemical weathering in action.

_____ 4. Since crystalline rocks are very hard and poor conductors of heat, they are not affected by diurnal heating and cooling processes.

_____ 5. With regard to weathering processes, sedimentary rocks are chemically very stable but are physically much weaker.

_____ 6. Hydration is the chemical weathering process whereby carbonate rocks such as limestone are dissolved.

_____ 7. Oxidation is the chemical combination of minerals with oxygen that has first been dissolved in water.

_____ 8. Chemical weathering processes are most active in polar and arid regions.

_____ 9. Physical weathering occurs most rapidly in regions that favor its primary mechanism of frost wedging.

_____ 10. Chemical and physical weathering are totally distinctive processes which occur in separate regions under different sets of environmental conditions.

_____ 11. The zone of weathering or "weathering front" is never more than one to three feet below the earth's surface.

_____ 12. The "angle of repose" is a theoretical measure of slope stability and is never encountered in the natural environment.

_____ 13. Mudflows move rapidly and are the most fluid of all of the types of mass movements.

_____ 14. Karst topography is formed by the solution weathering of limestone and dolomite.

_____ 15. Sinkholes, boulder fields, and exfoliation domes are characteristic landform features on a karst plain.

TESTING: MULTIPLE CHOICE

16. Exfoliation domes are produced:
     a. by chemical weathering
     b. in areas of crustal unloading
     c. by slump activity
     d. on karst plains

17. Talus cones are mainly found:
     a. in mountainous areas at the base of slopes
     b. in humid tropical regions
     c. in areas of solution weathering
     d. in areas of active soil creep

18. The absorption of water molecules within the molecule of
     a mineral is an example of:
     a. hydrolysis
     b. exfoliation
     c. oxidation
     d. hydration

19. The process of the dissolving of limestone and dolomite
     rocks is called:
     a. hydration
     b. oxidation
     c. carbonation
     d. hydrolysis

20. Karst topography is formed primarily by:
     a. carbonation and solution weathering
     b. mass wasting processes
     c. oxidation and hydrolysis
     d. exfoliation in combination with hydration

21. The main motivating force in mass wasting movement is:
     a. joint patterns in rocks
     b. slumps and earth flows
     c. solution weathering
     d. gravity

22. Mass movements of materials downslope are:
     a. always slow like slumping
     b. always rapid as in the case of mudflows and creep
     c. either rapid or quite slow
     d. always called avalanches

23. The steepest angle that can be maintained on any slope
     without failure of the slope materials is called the:
     a. angle of repose
     b. angle of cohesion
     c. angle of friction
     d. angle of exfoliation

24. Which of the following is <u>not</u> a rapid mass movement
     process:
     a. mudflow
     b. rockslide
     c. creep
     d. slump

25. Which of the following mass movements might be expected
     to occur in a mountainous volcanic region following a
     thunderstorm:
     a. slump
     b. mudflow
     c. creep
     d. subsidence sinkhole

26. Which of the following are the most widespread and best
     known of the major karst features:
     a. sinkholes
     b. uvalas
     c. exfoliation domes
     d. talus cones

27. Silver Springs, Florida is a classic example of a:
     a. cave
     b. natural bridge
     c. uvala
     d. sinking creek

28. Which of the following is <u>not</u> a major area of karst
     topography:
     a. Yucatan Peninsula, Mexico
     b. Central California
     c. Northwestern Yugoslavia
     d. Southern Florida

29. Examples of remnant hilly features on karst surfaces are:
     a. uvalas
     b. talus cones
     c. haystack hills and tower karst
     d. sinkholes and sinking creeks

TESTING: SHORT ANSWER AND ESSAY

1. In what major ways does weathering prepare surface rock
    materials for the process of gradation? How deeply does
    the weathering front extend into the bedrock?

2. What are the major forms of physical weathering and in
    which regions and under what environmental conditions do
    they predominate?

3. What are the four major types of chemical weathering? In
    what types of environments do these occur most notice-
    ably? Do these processes operate separately from the
    physical weathering processes mentioned earlier?

4. What is the connection between gravity and mass movements
    of materials downslope? Is this a true gradational
    process?

5. How do rapid and slow mass movements differ in terms of
    materials conveyed downslope and conditions which create
    them?

6. What is karst topography? Under what environmental
    circumstances does it occur? Could you list the main
    world areas of karst topography?

7. Why are areas of active solution weathering hazardous for
    human occupation? Are these same areas attractive ones
    for tourism? If so, what are the landform features which
    provide the basis for such an attraction?

8. What are the hazards of active mass movement of slope
    materials upon human occupation? Are there any safe-
    guards or precautions that might be taken to help
    mitigate these hazards?

9. What are the four main reasons for land subsidence? Why
    do these occur and what impacts does this occurrence have
    on human land use systems?

C H A P T E R   17

FLUVIAL PROCESSES AND LANDFORMS

LEARNING OBJECTIVES

1. Understand a basic premise of this chapter -- that streams and landforms formed by streams dominate landscapes through the world.

2. Survey the characteristics of the major rivers of the world and examine the type of solid load carried by these rivers to oceans and seas.

3. Introduce the pathway by which rainfall infiltrates the ground and is finally channelized into a flowing stream.

4. Discuss the role of vegetation is limiting stream action and erosion.

5. Examine the drainage basin as the fundamental unit of geomorphic study. Develop the terminology necessary to discuss the attributes of basins. Understand the basic geomorphic importance of streams.

6. Survey the concept of stream size, placing particular emphasis on stream order, stream patterns, and basin relationships.

7. Relate stream drainage patterns to the underlying materials.

8. Examine erosional processes, placing particular attention on the mechanisms by which streams transport materials.

9. Discuss the concept of grade and the flow of energy in a
    fluvial system.

10. Survey the types of landforms produced by streams,
    placing emphasis on distinguishing depositional and
    erosional features.

11. List the principal landforms developed by a meandering
    river in a floodplain; discuss the origin of these
    landforms.

12. Examine the various forms of the terminal landform
    created by a stream: the delta.

13. Closely examine the instability of the lower portion of
    the Mississippi River. Analyze the repercussions if a
    major shift in the current channels occurs. Examine the
    role of human intervention in stabilizing the channel.

KEY TERMS AND CONCEPTS

Provide a short definition or description of the following key
words and concepts from the chapter.

Basic Flow

Fluvial

Interfluve

Infiltration Capacity

Overland Flow

Rills

Gullies

Channelized Flow

Drainage System

Tributaries

Drainage Divides

Stream Order

Stream Pattern

Dendritic Pattern

Trellis Pattern

Radial Pattern

Centripetal Pattern

Annular Pattern

Rectangular Pattern

Deranged Pattern

Erosional Processes

Critical Erosion Velocity

Stream Banks

Stream Bed

Turbulence

Hydraulic Action

Solution

Abrasion

Stream Load

Dissolved Load

Suspension Load

Saltation Load

Bed Load

Base Level

Graded Stream

Floodplain

Stream Geometry and Plan

      Meandering

      Straight

      Braided

Stream Geometry and Plan

      Longitudinal Profile

      Knickpoints

      Graded

Stream Sediments

      Sorted

      Graded

      Alluvium

Fluvial Landforms

      Valley

      Terraces

      Gorge and Canyon

      Floodplains

      Point Bars

      Meander Neck

      Oxbow Lake

      Natural Levee

Deltas

      Bird's Foot Delta

      Classical Triangular

      Estuaries

      Distributary Channels

TESTING: TRUE/FALSE

_____   1. An erosional hill is a highland remnant that remains after dissection of a plateau.

_____   2. The Atchafalaya channel is located in Mississippi.

_____   3. Rivers make good borders because they are stable in position over long time periods.

_____   4. The material found in a bar along a river will typically be sand or gravel.

_____   5. The change in the gradient of a stream along its course is called the longitudinal profile.

_____   6. An abrupt increase in the slope of a stream is called a knickpoint.

_____   7. The most widely distributed drainage pattern is the trellis pattern.

_____   8. A small shoestring channel is called a rill.

_____   9. The principal mechanism of carrying solid material in a stream is in suspension.

_____   10. Floodplain formation is initiated while down-cutting of the channel is still active.

_____   11. Stability of the lower Mississippi has been at the price of a threatened catastrophic shift in the channel.

_____   12. A tributary stream is the type of stream which is generally found in deltas.

_____   13. A broad looping bend in a river is called a meander.

_____   14. When sediment is being transported, the minimum power of water required to initiate movement of a particle is the critical erosion velocity.

_____   15. The initial flow of rainfall over the land surface is unlikely to be in channels; it is called over-land flow.

_____   16. Stream patterns evolve independently of the under-lying materials and geologic structures.

_____ 17. Three-fourths of the earth's land surface is dominated by landforms formed by fluvial processes.

_____ 18. The average erosion of the Mississippi River is
five centimeters per 1000 years.

_____ 19. Friction is generated along the banks and the
bottom (bed) of rivers.

_____ 20. Discharge computations require velocity and
channel area measurements.

TESTING: MULTIPLE CHOICE

Choose the best response for each of the following multiple
choice question. Each question has only one correct answer.

21. The coarsest particles which can be carried by a stream
are carried as:
a. bed load
b. saltation load
c. suspended load
d. dissolved load

22. The basic mechanisms by which fluvial landforms are
formed are:
a. erosion and deposition
b. solution and corrosion
c. corrasion and corrosion
d. impact and erosion

23. The gorge of the Colorado River:
a. is an example of solution
b. is over 1.6 kilometers deep
c. developed during the glacial period
d. is an example of a fault block

24. Floodplains:
a. are infrequently encountered in fluvial landscapes
b. are never depositional in origin
c. often evolve after a stream has ceased downcutting of
its bed
d. are the areas of a stream present as terraces; they
are no longer subject to flooding

25. A bird's-foot delta is found at the terminus of:
a. the Nile River
b. the Ganges River
c. the Niger River
d. the Mississippi River

26. A semi-enclosed body of water in a drowned river valley
    is:
    a. called a lake
    b. called estuary
    c. associated with volcanic craters
    d. the result of slow subsidence of deltaic material

27. The proportion of the United States drained by the
    Mississippi River is:
    a. 25%
    b. 40%
    c. 55%
    d. 70%

28. Which is not one of the two vital natural functions of a
    stream?
    a. removal of excess water from the land
    b. generation of fluvial landforms
    c. removal of surface weathering products from the land
    d. each of the above answers is a vital function of a
        stream

29. The progressive narrowing of a meander loop does not
    results in:
    a. a point bar
    b. an oxbow lake
    c. a meander neck
    d. a meander cutoff

30. Which of the following is not an indirect result of
    stream erosion?
    a. mountains and hills
    b. point bars
    c. deltas
    d. gorges

31. Which of the following is not a measure of stream size?
    a. basin area and stream order
    b. stream length and width
    c. stream discharge
    d. each of the above answers is a measure of stream size

32. The landscapes in which flowing water is the most
    effective:
    a. tropical landscapes
    b. semi-arid landscapes
    c. humid landscapes
    d. polar landscapes

33. Which is not a traditional use of streams by human
     society?
     a. transportation
     b. irrigation
     c. desiccation
     d. agriculture on the floodplains

34. The movement of material in a molecular or atomic state:
     a. is called solution
     b. is called traction
     c. is called electrolytic movement
     d. is called saltation

35. The type of drainage pattern associated with a volcano:
     a. is dendritic
     b. is parallel
     c. is radial
     d. is rectangular

1. Make a listing of the major types of drainage patterns
    and the types of landforms they are associated with.

2. Describe the four principal means by which streams move
    solid materials to the oceans. With what particle sizes
    is each mechanism associated?

3. Name the major types of river deltas and give an example
    of each.

4. Describe the concept of a graded stream.

5. Why is the critical erosional velocity of a particle
    greater than the transportation velocity?

6. Name the basic means of measuring the size of a stream.
    Which measure do you regard to be most useful? Why?

7. Why do mature alluvial rivers make insecure political
    boundaries?

8. What are some of the difficulties encountered in the
    attempt to control the flow and the channel of a river
    such as the lower Mississippi?

C H A P T E R   18

GLACIAL AND PERIGLACIAL LANDFORMS

1. Identify the two major types of glaciers and outline the conditions under which they are formed.

2. Describe the erosional work done by both continental ice sheets and alpine glaciers.

3. Discuss the impact of alpine glaciation in sculpturing the topography of many of the earth's mountainous regions and identify the main areas which have been so modified.

4. Define the Pleistocene Epoch as a unique geological period in the earth's recent history and discuss the salient climatic events which were involved as well as the portions of the earth's continents which were affected.

5. Compare and contrast the erosional and depositional work accomplished on the three major plains associated with Pleistocene continental glaciers: ice-scoured plains, till plains, and outwash plains.

6. Relate the interrelationships that exist between the three environmental phenomena of glaciers, climate, and the earth's ocean levels.

7. Discuss the types of environmental conditions that exist and create landforms in periglacial regions both during the Pleistocene Epoch and at the present time in the subpolar and alpine areas of the earth.

8. Outline some of the major theories which have been
   advanced to account for the type of climatic change that
   encouraged large-scale glaciation to occur during the
   Pleistocene.

KEY TERMS AND CONCEPTS

Pleistocene Epoch

Glacier

Continental Glacier

> Zone of Accumulation
>
> Zone of Ablation
>
> Ice-Scoured Plain
>
> Laurentide Ice Sheet
>
> Striations
>
> Roche Moutonees
>
> Glacial Drift
>
> Glacio-Fluvial Sediments
>
> Moraines
>
>> Terminal Moraine
>>
>> Ground Moraine
>>
>> Recessional Moraine
>
> Outwash Plain

<u>Alpine Glacier</u>

> Cirque
>
> Crevasse
>
> Lateral Moraine
>
> Medial Moraine
>
> Tarn
>
> Arete
>
> Horn
>
> Glacial Trough
>
> Valley Train
>
> Kame
>
> Hanging Valley
>
> Fjord

<u>Continental Glacier Depositional Landforms</u>

> Till Plain
>
> Kettle
>
> Drumlin
>
> Outwash Plain
>
> Glacial Lacustrine Plain
>
> Esker

Periglacial

Permafrost

Patterned Ground

Solifluction

_____   1. Continental glaciation occurred only during the Pleistocene but alpine glaciation still occurs today.

_____   2. In an active glacier, the zone of ablation is the area where snow is accumulating and being compacted into ice.

_____   3. Striations are gouges that rocky cliffs carve into the passing mass of glacial ice.

_____   4. Glacial drift is the general term for all deposits of glacial origin.

_____   5. Glacial till is unsorted debris directly deposited by a glacier, in contrast to glacio-fluvial sediment which is sorted material deposited by glacial meltwaters.

_____   6. A recessional moraine is a mound of glacial till that marks the furthest point of advance of a glacier.

_____   7. The heads of most alpine glaciers are contained in bowl-shaped depressions called cirques.

_____   8. A tarn is a small lake that is found in a cirque which is today devoid of ice.

_____   9. A narrow, serrated ridge created by headward cirque action on several sides is called a fjord.

_____   10. The Pleistocene Epoch began about two million years ago and is still continuing.

_____   11. During the Pleistocene, glacial ice covered about 85 percent of the earth's present land area.

_____   12. Drumlins are smooth, elongated hills found on the till plain of a former continental glacier.

_____   13. Eskers and kettles are erosional features of glacio-fluvial origin found on the outwash plain.

_____   14. The laurentian shield of Eastern Canada is a major ice-scoured plain which was created during the Pleistocene.

_____   15. The movement of material downhill in periglacial regions is called solifluction.

TESTING: MATCHING

(Letters may be used more than once)

A. EROSIONAL FEATURE:
      ALPINE GLACIATION

B. DEPOSITIONAL FEATURE:
      ALPINE GLACIATION

C. EROSIONAL FEATURE:
      ICE-SCOURED PLAIN

D. FEATURE ON THE TILL
      PLAIN

E. ASSOCIATED WITH THE
      OUTWASH PLAIN

F. ASSOCIATED WITH
      PERIGLACIAL ACTIVITY

_____ 16. Solifluction

_____ 17. Drumlin

_____ 18. Rock Basin

_____ 19. Roche Moutonee

_____ 20. Esker

_____ 21. Cirque

_____ 22. Ground Moraine

_____ 23. Tarn

_____ 24. Lateral Moraine

_____ 25. Arete

_____ 26. Medial Moraine

_____ 27. Patterned Ground

_____ 28. Horn

_____ 29. Kettle

_____ 30. Fjord

TESTING: MULTIPLE CHOICE

31. The loss of glacial ice by melting and sublimation is
     termed:
     a. regelation
     b. crevasse
     c. ablation
     d. firn

32. A small, rounded bedrock feature caused by glacial
     erosion is a:
     a. roche moutonee
     b. striation
     c. moraine
     d. drumlin

33. Which of the following is not material directly deposited
     by glacial ice:
     a. recessional moraine
     b. valley train
     c. ground moraine
     d. glacial till

34. A tarn is a lake which is found today in a:
     a. till plain
     b. horn
     c. glacial trough
     d. cirque

35. Which of the following is not associated with
     Switzerland's Matterhorn:
     a. arete
     b. cirque
     c. kame
     d. horn

36. A glacial trough which becomes drowned by seawater is
     called a:
     a. fjord
     b. kame
     c. cirque
     d. tarn

37. Which of the following is not an erosional feature
     associated with an ice-scoured plain:
     a. rock basin lakes
     b. soil removal
     c. roche moutonees
     d. valley train

38. Which of the following is <u>not</u> a depositional feature
     found on a glacial till plain:
     a. ground moraine
     b. drumlin
     c. roche moutonee
     d. esker

39. Which of the following is <u>not</u> associated with a glacial
     outwash plain:
     a. recessional moraine
     b. kettle
     c. stratified material
     d. glacio-fluvial sediments

40. Eskers are:
     a. low in elevation than the surrounding landscape
     b. long, meandering ridges of stratified drift
     c. found on the ice-scoured plain
     d. also called drumlins

41. Which of the following is <u>not</u> associated with periglacial
     activity:
     a. permafrost
     b. patterned ground
     c. kettles
     d. solifluction

42. Recent studies of the growth patterns of Alaskan glaciers
     have shown that:
     a. all are surging
     b. all are retreating
     c. each glacier either surges or retreats
     d. some are surging and some are retreating

43. The Pleistocene Epoch lasted about:
     a. 10,000 years
     b. 500,000 years
     c. 2 million years
     d. 10 million years

<u>TESTING: SHORT ANSWER AND ESSAY</u>

1. What are the two major types of glaciers? What are the environmental conditions under which they form?

2. What is the overall erosional and depositional work that is accomplished by alpine glaciers? Which of the earth's mountain systems can be considered to have been glaciated?

3. What was the Pleistocene Epoch? What are some of the hypotheses for the origin of the Pleistocene? Is there evidence that this period is over?

4.   Regarding continental glaciation, what are the erosional
     features that glaciers leave on ice-scoured plains?
     Provide examples of some major world regions that have
     been affected in this way?

5.   How do the landforms found on glacial till plains differ
     from the landforms found on outwash plains? In what ways
     is the overall topography and the conditions that created
     it different over these two plains?

6.   Under what conditions do periglacial conditions occur and
     what are some of the landforms that are created? What
     role does permafrost play in these processes? What world
     regions are presently affected? Was there a different
     pattern of periglacial regions during the Pleistocene?

7. How does the relationship between the Spokane Flood and
    the Channelled Scablands relate to the geological
    principle of "uniformitarianism"? What considerations
    are involved?

8. Discuss the relationships between changes in the earth's
    orbital geometry and climatic changes over the earth.
    How could ice ages such as the Pleistocene be related to
    this theory?

EOLIAN PROCESSES AND DESERT LANDSCAPES

LEARNING OBJECTIVES

1. Discuss the relative importance of fluvial and eolian processes in desert landscapes.

2. Identify the major types of sand dunes and relate them to the prevailing wind directions.

3. Examine the expansion of desert areas and the effect this has on the peoples living in these marginal regions.

4. Develop the terminology which is used to name the types of unique landforms of desert regions of the world.

5. Identify why deserts are so remarkably different in appearance from humid areas, even though fluvial processes dominate each landscape.

6. Examine loess deposits, an under-acknowledged feature resulting from eolian transport of silt grains for great distances.

7. Examine the drainage characteristics of deserts.

8. Understand the expansion of desertic conditions worldwide in the process known as desertification.

9. Gain an overall understanding of the various types of landforms found on desert surfaces.

KEY TERMS AND CONCEPTS

Provide a short definition or description of the following key words and concepts from the chapter.

Wind Processes

    Eolian

    Drag Velocity

    Surface Creep

    Saltation

    Dust Storm

    Sandstorm

Wind Erosion

    Abrasion

    Deflation

    Ventifacts

    Deflation Hollows

    Blowouts

    Desert Pavement

Wind Deposited Landforms

    Loess

    Dunes

        Back Slope

        Slip Face

    Erg

    Barchan

    Longitudinal Dune

    Transverse Dune

Star Dune

Sand Sheets

Sand Shadows

Sand Ripples

Landforms

Desert

Hamada

Exotic River

Interior Drainage

Back-Weathering

Pediment

Bolson

Playa

Alluvial Fan

Peidmont Alluvial Fan

Washes (Arroyos)

Back-wasting

Duricrust Layer

Mesa

Butte

Bajada

Pediplain

Inselberg

Desertification

TESTING: TRUE/FALSE

_____   1.   The origin of the term for wind geomorphic pro-
                cesses goes bask to the Greek god Aeolus.

_____   2.   The erosive effect of the wind (due to the
                sandblasting effect) on consolidated materials
                is very large.

_____   3.   Loess is airborne silt particles carried by the
                wind as suspended load.

_____   4.   In general, the regime of desert streams is
                ephemeral.

_____   5.   A resistant rock layer which protects the
                underlying rock from erosion is a cap rock.

_____   6.   A duricrust is a critical component of island
                mountains called inselbergs.

_____   7.   Desertification is accomplished through the loess
                of stabilizing vegetation followed by accelerated
                erosion by wind and water.

_____   8.   Livestock raising is unimportant when considering
                the causes and solutions for desertification.

_____   9.   A ventifact is unlikely to be found on a desert
                pavement.

_____   10.   The slip face of a sand dune is less steep than
                the back slope.

_____   11.   The intensity of use of marginal lands should not
                exceed their carrying capacity in the best years.

_____   12.   There is no difference between a pediplain and a
                peneplain other then their respective occurrences
                in humid and arid areas.

_____   13.   An alluvial fan is a common landform which forms
                when a desert stream exits from mountains to a
                plain.

_____   14.   A hamada is a subclassification of desert land-
                forms consisting of bare rock surfaces.

_____   15.   The most common and most familiar type of sand
                dune is an erg.

_____ 16. Stabilized sand dunes are found where vegetation is able to maintain a cover. Prominent examples are in Western Nebraska and in the Sahara.

_____ 17. The impact of saltating sand grains results in abrasion.

_____ 18. The underlying factor controlling the differences between desert and non-desert regions is the availability of surface moisture.

_____ 19. The Nile River is an example of an exotic river.

_____ 20. An outlying residual with a flat top is called a mesa.

TESTING: MULTIPLE CHOICE

Choose the best response for each of the following multiple choice question. Each question has only one correct answer.

21. Which is not a type of sand dune?
    a. erg
    b. longitudinal
    c. hamada
    d. barchan

22. The humid climate equivalent of an inselberg is a:
    a. mesa
    b. butte
    c. duricrust
    d. monadnock

23. Desertification:
    a. results in human suffering, hardship, and death
    b. is restricted to Australia
    c. results from climate change, not from human action
    d. was brought to global attention by the Sahelian drought in South America

24. Loess:
    a. is wind deposited sand materials
    b. blankets much of the Midwest of the United States
    c. is seldom more than a few inches thick
    d. is generally found in desert areas

25. Which of the following is not the name of a common small
    fluvial landform?
    a. wash
    b. gorge
    c. wadis
    d. arroyo

26. In general:
    a. longitudinal dunes are perpendicular to prevailing
       winds
    b. transverse dunes are parallel to prevailing winds
    c. barchans are parallel to prevailing winds
    d. star dunes form in areas of variable wind

27. A sand ripple:
    a. is similar in appearance to a water ripple
    b. can grow to several feet in height
    c. have a long life span in deserts
    d. grow parallel to the wind

28. Loess:
    a. forms distinctive high-relief landforms
    b. travels by saltation
    c. is associated with featureless plains
    d. cannot be used for agriculture

29. Deserts differ from humid areas:
    a. because fluvial action is unimportant in deserts
    b. because of the angularity of topography
    c. because sandy soils do not occur in humid areas
    d. because deserts cannot occur in cold climates

30. Which is associated with deserts?
    a. back-wearing of slopes
    b. down-wearing of slopes
    c. diagonal stripping of slopes
    d. a high water table

TESTING: SHORT ANSWER AND ESSAY

1.   List the main causes and consequences of desertification.

2.   What is the difference between a pediment and a pediplain?

3.   Name the principal processes which result in deserts having a distinctly different appearance than the landscapes of the humid middle latitudes.

4.   What are the differences among abrasion, deflation, and
     saltation?

5.   Make a table listing the major types of sand dunes and
     the direction of the prevailing wind with which they are
     associated.

6. Make a list of the major deserts of the world, include
    the continent and the latitude at which they are found.
    What climatic and topographic factors combine to produce
    deserts?

7. Provide a cross-section of a desert basin showing the
    landform features on both the slopes and on the bolson.
    Be careful to indicate the drainage systems involved.

C H A P T E R   20

COASTAL LANDFORMS AND SEAFLOOR TOPOGRAPHY

LEARNING OBJECTIVES

1. Develop a familiarity with the names of coastal landforms.

2. Examine the unique forces associated with wave action on the bedrock materials of coasts.

3. Present coastlines as landforms related to changes in sea level; characteristic forms develop along emergent and along submergent coastlines.

4. Examine the plausibility of extracting large quantities or mineral deposits from the floor of the oceans.

5. List the picturesque landforms found along coastlines and provide explanations for their genesis. A full appreciation of the geomorphology of coasts requires an understanding of how the landscapes evolve.

6. Investigate the hidden landforms of the seafloor; analyze their origin and relate them to tectonic processes.

7. Focus on the beach as the basic unit of study for coastal landforms -- this is the most active environment and is found in all coastal areas.

8. Become familiar with the main types of coasts as characterized by the advancement or retreat of the coastal shoreline.

KEY TERMS AND CONCEPTS

Provide a short definition or description of the following key words and concepts from the chapter.

Beach Processes

    Abrasion

    Attrition

    Swash

    Backwash

    Solution

    Hydraulic Action

Beach Movements and Forms

    Beach Drifting

    Longshore Currents

    Wave Refraction

    Headlands

    Undertow

    Rip Currents

    Sea Cliff

    Sea Arch

    Stack

    Pocket Beach

    Wave-Cut Platform

    Wave-Built Platform

    Beach

    Sand Bar

    Barrier Island

Spit

Tidal Inlet

Lagoon

Mudflat

Salt Marsh

Coastal Types

Retreating Coasts

Erosional Coasts

Submergent Coast

Ria Coast

Emergent Coasts

Fjords

Marine Terrace

Advancing Coasts

Depositional Coasts

Volcanic Coasts

Coral Coasts

Atoll

Fringing Reef

Barrier Reef

Landforms and Processes of the Ocean Floor

Continental Shelf

Submarine Canyon

Turbidity Currents

Continental Slope

Abyssal Plain

Seamount

TESTING: TRUE/FALSE

_____   1. Two processes associated with wave action are
               abrasion and attrition.

_____   2. Longshore currents are most common where the wave
               action is perpendicular to the coast.

_____   3. The shoreward flow of water in breaking waves is
               balanced by the outward flow of the undertow.

_____   4. Sea cliffs are produced by strong wave action on
               weak bedrock materials.

_____   5. The types of materials in wave-built platforms can
               include mud, sand, gravel, and cobbles.

_____   6. A lagoon is most likely to form behind a barrier
               island and be connected to the sea by a tidal
               inlet.

_____   7. Rapid erosion of coasts is associated with
               emergent coastlines.

_____   8. The main difference between an atoll and either a
               fringing reef or a barrier reef is that an atoll
               is not associated with a larger land mass.

_____   9. Tubidity currents are a major process in
               subsurface oceanic landscapes close to coasts.

_____   10. Separating the deep seafloor from the continental
               shelf is a transitional area called the
               continental slope.

_____   11. Both ria coasts and fjord coasts contain deep
               indentations.

_____   12. Most coastlines are presently retreating.

_____   13. Marine terraces are characteristic along
               submergent coasts.

TESTING: MULTIPLE CHOICE

Choose the best response for each of the following multiple
choice question.   Each question has only one correct answer.

14.   Each of the following landforms occurs in a similar type
      of situation except:
      a.   marine terrace
      b.   coral atoll
      c.   lagoon
      d.   fringing reef

15.   When selecting a beach for general lounging purposes it
      would be best to go to:
      a.   an emergent coastline
      b.   a deltaic coastline
      c.   a shore bordered by a cliff
      d.   a barrier beach

16.   Coastlines advance seaward because of:
      a.   tectonic or isostatic uplift
      b.   downwarping of the coast
      c.   melting of glacial waters
      d.   submergence following the Pleistocene

17.   Barrier beaches:
      a.   are stable sites for human occupancy
      b.   have a long life span
      c.   provide protected ports
      d.   are unsuited for towns and cities

18.   Geysers:
      a.   occur on emergent headlands
      b.   are associated with wave destruction of cliffs
      c.   can occur on the deep ocean floor
      d.   none of the above answers is correct

19.   Fjord coasts:
      a.   occur in high latitudes
      b.   correlate with glacial valleys
      c.   are deeply indented
      d.   all of the above

20.   Barrier beaches:
      a.   never migrate
      b.   can be destabilized by human activities
      c.   do not occur in the United States
      d.   are found around atolls

TESTING: SHORT ANSWER AND ESSAY

1. Describe alternatives which might protect a town developed on a barrier beach.

2. Describe the atmospheric, terrestrial, and marine processes which combine to produce the landforms of coastal areas.

3. Outline the progressive history of a submergent coastline. Where are these coasts found in the United States.

4. What is a beach? What processes combine to create beach
    deposits? Can beaches be protected from erosion?

5. What are the processes which operate along <u>coral coasts</u>?
    How are atolls formed? Where do most atolls occur today?

6. What are marine terraces and how are they formed? Where
    might these coastal landforms be found in the United
    States?

CHAPTER ONE

True/False		Multiple Choice	
1.	F	21.	d
2.	T	22.	c
3.	F	23.	a
4.	F	24.	d
5.	F	25.	b
6.	T	26.	d
7.	T	27.	d
8.	T	28.	a
9.	F	29.	d
10.	F	30.	c
11.	T	31.	d
12.	F	32.	b
13.	T	33.	d
14.	F		
15.	F		
16.	F		
17.	T		
18.	F		
19.	T		
20.	T		

CHAPTER TWO

Matching I

1. F
2. E
3. G
4. B
5. D
6. H
7. J
8. C
9. I
10. A

Matching II

11. D
12. F
13. C
14. J
15. A
16. B
17. H
18. E
19. I
20. G

Multiple Choice

21. a
22. c
23. b
24. b
25. c
26. c
27. a
28. b
29. c
30. a
31. b
32. d
33. c
34. a
35. a

CHAPTER THREE

<u>Matching</u>   <u>Multiple Choice</u>

1.	F		11.	d
2.	F		12.	a
3.	T		13.	d
4.	T		14.	d
5.	T		15.	b
6.	T		16.	c
7.	F		17.	a
8.	F		18.	b
9.	F		19.	c
10.	T		20.	a
			21.	d
			22.	c
			23.	b
			24.	a
			25.	d

CHAPTER FOUR

Matching

1. I
2. M
3. H
4. F
5. B
6. D
7. A
8. N
9. K
10. L
11. C
12. E

Multiple Choice

13. b
14. d
15. a
16. b
17. b
18. a
19. d
20. b
21. d
22. d
23. d
24. a
25. d
26. d
27. c
28. c
29. c
30. e

CHAPTER FIVE

Matching   Multiple Choice

    1.   I   13.   c
    2.   G   14.   d
    3.   M   15.   a
    4.   H   16.   b
    5.   E   17.   d
    6.   A   18.   c
    7.   C   19.   d
    8.   B   20.   a
    9.   N   21.   d
   10.   J   22.   a
   11.   F   23.   c
   12.   D   24.   c
                                        25.   b
                                        26.   a
                                        27.   d
                                        28.   b
                                        29.   c
                                        30.   c
                                        31.   b
                                        32.   a

CHAPTER SIX

True/False

1.	F		6.	F
2.	F		7.	T
3.	F		8.	T
4.	T		9.	T
5.	T		10.	T

Matching

11.	H		16.	B
12.	J		17.	I
13.	C		18.	D
14.	E		19.	F
15.	G		20.	A
			21.	K

Multiple Choice

22.	a		29.	a
23.	a		30.	b
24.	a		31.	d
25.	c		32.	d
26.	b		33.	a
27.	c		34.	a
28.	b		35.	d
			36.	d

CHAPTER SEVEN

Matching I

1. C
2. H
3. D
4. B
5. I

6. G
7. A
8. J
9. E
10. F

Matching II

11. D
12. B
13. H
14. I
15. C

16. G
17. A
18. F
19. J
20. E

Multiple Choice

21. c
22. d
23. b
24. a
25. d
26. c
27. a
28. b

29. d
30. b
31. a
32. a
33. c
34. c
35. a
36. c

CHAPTER EIGHT

True/False

1.	F		7.	T
2.	T		8.	F
3.	T		9.	F
4.	F		10.	T
5.	F		11.	F
6.	F		12.	T

Matching

13.	I		19.	J
14.	O		20.	D
15.	A		21.	M
16.	G		22.	K
17.	H		23.	E
18.	L		24.	B

Multiple Choice

25.	a		31.	a
26.	d		32.	d
27.	b		33.	b
28.	d		34.	c
29.	a		35.	a
30.	b		36.	d

CHAPTER NINE

True/False

1.	F		11.	T
2.	T		12.	F
3.	F		13.	F
4.	F		14.	T
5.	F		15.	T
6.	F		16.	F
7.	T		17.	T
8.	F		18.	F
9.	T		19.	F
10.	T		20.	T

Matching I

21.	H		27.	L
22.	J		28.	K
23.	F		29.	C
24.	G		30.	D
25.	A		31.	B
26.	I			

Matching II

32.	D		40.	C
33.	E		41.	B
34.	C		42.	D
35.	F		43.	G
36.	G		44.	L
37.	A		45.	I
38.	J		46.	H
39.	A			

Multiple Choice

47.	c		56.	b
48.	b		57.	a
49.	a		58.	a
50.	d		59.	b
51.	d		60.	c
52.	a		61.	d
53.	a		62.	c
54.	d		63.	d
55.	d			

CHAPTER TEN

True/False

1. T	7. F
2. T	8. T
3. F	9. F
4. F	10. F
5. T	11. T
6. F	12. F

Matching

13. G	18. J
14. D	19. A
15. I	20. C
16. H	21. F
17. E	22. B

Multiple Choice

23. a	30. b
24. a	31. b
25. b	32. c
26. c	33. d
27. b	34. b
28. a	35. a
29. b	36. c
	37. a

CHAPTER ELEVEN

True/False

1. F	11. T
2. T	12. F
3. F	13. F
4. F	14. T
5. F	15. F
6. T	16. F
7. T	17. T
8. T	18. F
9. T	19. T
10. F	20. T

Matching

21. D	27. I
22. L	28. C
23. J	29. N
24. E	30. B
25. M	31. F
26. A	32. G

Multiple Choice

33. c	39. d
34. b	40. b
35. a	41. b
36. d	42. c
37. c	43. b
38. a	44. b

CHAPTER TWELVE

True/False

1.	T	6.	T
2.	F	7.	F
3.	T	8.	T
4.	F	9.	F
5.	T	10.	T

Matching

11.	I	16.	G
12.	D	17.	H
13.	F	18.	J
14.	C	19.	E
15.	B	20.	A

Multiple Choice

21.	b	28.	b
22.	a	29.	b
23.	d	30.	c
24.	b	31.	d
25.	a	32.	a
26.	c	33.	c
27.	a	34.	a
		35.	d

CHAPTER THIRTEEN

True/False

1.	T		6.	T
2.	F		7.	F
3.	F		8.	F
4.	T		9.	T
5.	T		10.	F

Multiple Choice

11.	b		32.	c
12.	a		33.	b
13.	d		34.	d
14.	d		35.	b
15.	a		36.	d

CHAPTER FOURTEEN

True/False

1.	F		8.	F
2.	F		9.	T
3.	T		10.	T
4.	F		11.	T
5.	T		12.	T
6.	T		13.	F
7.	F		14.	T

Matching

15.	C		20.	C
16.	F		21.	F
17.	A		22.	D
18.	D		23.	B
19.	C		24.	E

Multiple Choice

25.	b		34.	d
26.	b		35.	b
27.	b		36.	a
28.	a		37.	d
29.	c		38.	b
30.	c		39.	d
31.	a		40.	b
32.	b		41.	c
33.	a		42.	d

CHAPTER FIFTEEN

True/False

1.	F		10.	T
2.	T		11.	F
3.	T		12.	T
4.	F		13.	T
5.	F		14.	F
6.	T		15.	F
7.	F		16.	T
8.	F		17.	T
9.	T			

Matching

18.	E		26.	D
19.	M		27.	O
20.	R		28.	Q
21.	I		29.	G
22.	S		30.	L
23.	T		31.	J
24.	A		32.	F
25.	H		33.	B

Multiple Choice

34.	b		42.	a
35.	d		43.	c
36.	d		44.	b
37.	a		45.	a
38.	c		46.	b
39.	a		47.	c
40.	b		48.	d
41.	d		49.	a
			50.	a
			51.	d

CHAPTER SIXTEEN

<u>True/False</u>

1.	T		9.	T
2.	T		10.	F
3.	F		11.	F
4.	F		12.	F
5.	T		13.	T
6.	F		14.	T
7.	T		15.	F
8.	F			

<u>Multiple Choice</u>

16.	b		23.	a
17.	a		24.	c
18.	a		25.	b
19.	c		26.	a
20.	a		27.	d
21.	d		28.	b
22.	c		29.	c

CHAPTER SEVENTEEN

True/False

1.	T		11.	T
2.	F		12.	F
3.	F		13.	T
4.	T		14.	T
5.	T		15.	T
6.	T		16.	F
7.	F		17.	T
8.	T		18.	F
9.	T		19.	T
10.	F		20.	T

Multiple Choice

21.	a		28.	b
22.	a		29.	a
23.	b		30.	c
24.	c		31.	d
25.	d		32.	b
26.	b		33.	c
27.	b		34.	a
			35.	c

CHAPTER EIGHTEEN

True/False

1.	F	8.	T
2.	F	9.	F
3.	F	10.	T
4.	T	11.	F
5.	T	12.	T
6.	F	13.	F
7.	T	14.	T
		15.	T

Matching

16.	F	24.	B
17.	D	25.	A
18.	C	26.	B
19.	C or A	27.	F
20.	D	28.	A
21.	A	29.	E
22.	D	30.	A
23.	A		

Multiple Choice

31.	c	37.	d
32.	a	38.	c
33.	b	39.	a
34.	d	40.	b
35.	c	41.	c
36.	a	42.	d
		43.	c

CHAPTER NINETEEN

True/False

1.	T	11.	F
2.	F	12.	F
3.	T	13.	T
4.	T	14.	T
5.	T	15.	F
6.	F	16.	T
7.	T	17.	T
8.	F	18.	T
9.	F	19.	T
10.	F	20.	T

Multiple Choice

21.	c	26.	d
22.	d	27.	a
23.	a	28.	c
24.	b	29.	b
25.	b	30.	a

CHAPTER TWENTY

True/False

1. T
2. F
3. T
4. F
5. T
6. T

7. F
8. T
9. T
10. T
11. T
12. T
13. F

Multiple Choice

14. a
15. d
16. a

17. d
18. c
19. d
20. d